A. Raouf
and
B.S. Dhillon

SAFETY
ASSESSMENT

A Quantitative Approach

LEWIS PUBLISHERS
Boca Raton   Ann Arbor   London   Tokyo

Library of Congress Cataloging-in-Publication Data

Raouf, A. (Abdul), 1929—
    Safety assessment : a quantitative approach / by A. Raouf, B. S.
Dhillon.
        p.   cm.
    Includes bibliographical references and index.
    ISBN 0-87371-675-2
    1. Industrial safety.   I. Dhillon, B. S.   II. Title.
    T55.R27   1993
    658.3'82--dc20   93-24946
                                                                                        CIP

© 1994 by CRC Press, Inc.
Lewis Publishers is an imprint of CRC Press

No claim to original U.S. Government works
International Standard Book Number 0-87371-675-2
Library of Congress Card Number 93-24946
Printed in the United States of America 1 2 3 4 5 6 7 8 9 0
Printed on acid-free paper

The program mentioned in Chapter 6 can be obtained free of charge from the authors of the book. The disks are by no means an essential part of the book, which can be utilized to a full extent without the disks.

Professor A. Raouf (author of the computer program)
Systems Engineering
King Fahd University of Petroleum & Minerals
Dhahran 31261
SAUDI ARABIA

Professor B.S. Dhillon
Director of Engineering Management Program
University of Ottawa
Faculty of Engineering
770 King Edward
Ottawa, Ontario K1N 6N5
CANADA

PREFACE

At present, none of the theories of accident causation are universally accepted. All such theories, regardless of their origin, are basically conceptual in nature and are of limited utility in improving safety performance. Therefore, due to the lack of a theoretical base, safety is not recognized as Science. Improving safety, as usually practiced, involves a critical examination of the worker, the work, and the environment.

A continuous improvement of safety performance is one of the desired goals of a successful organization. To be able to embark on a continuous improvement program, it is necessary that safety performance be quantified. It is with this intention that the authors have written this text. In general, management and engineering aspects of safety are not covered in the text, as there are many excellent publications already available in these areas. In any case, the readers will find such material is not a prerequisite to understanding the content of this text. This book can be used as a text to teach senior undergraduate and graduate level students as well as a reference for professionals concerned with safety.

Chapters 1 and 2 introduce the subject and point out the need for a safety criterion. Commonly used measures of safety performance are introduced and explained. Chapter 3 is directed to those who have taken a course in Probability and Statistics. It may also be useful for those who have already studied Probability and Statistics, as the material presented is slanted towards accidents and safety performance. In Chapter 4, safety behaviour sampling is introduced and a procedure for safety behaviour sampling along with the application of control charts is described. Chapter 5 discusses the method of conducting safety appraisals and drawing statistical inferences. In Chapter 6, Safety Information System (SIS) is described and a PC-based SIS is explained. Chapter 7 discusses automation and safety related topics. The sources of automation related hazards are discussed. The appendix presents Poisson distribution, standard normal distribution and t distribution tables, a list of unsafe acts. safety inspection checklist, SIS description and the user manual, and selected codes from ANSI Z 16.2.

---

* The dBase-compatible computer program can be obtained free of charge from the authors, whose addresses are in the front of the book.

The authors sincerely wish to thank Longman Group Ltd, John Wiley & Sons, Macmillan Publishing Company, Garland Publishing, Inc., and Professor B. Lindgren of the University of Minnesota for granting permission to use their copyrighted material. Furthermore, the authors are grateful to the American National Standards Institute, Inc. for the use of selected codes from ANSI Z to 16.2 even though authors were told that this document is withdrawn. All in all, to the best of our knowledge we have obtained permissions from all concerned people to use their material. If inadvertently we are in error, the authors sincerely apologize to those whom we overlooked.

The assistance provided by the King Fahd University of Petroleum and Minerals to the first author is gratefully appreciated.

A. Raouf
Dhahran

B. S. Dhillon
Ottawa

DEDICATION

This book is affectionately dedicated to the memory of Dr. H. M. Yusaf and Ganda S. Dhillon.

THE AUTHORS

Dr. A. Raouf joined the Department of Systems Engineering, King Fahd University of Petroleum and Minerals, Dhahran as a Professor in 1984. Currently, he is the Chairman of this department. Before joining the King Fahd University of Petroleum and Minerals, he was a faculty member in the Department of Industrial Engineering at the University of Windsor, Ontario, Canada. During 1974 to 1983, he was the Head of the Industrial Engineering Department at the University of Windsor. He is a registered professional engineer in the Province of Ontario. He is actively engaged in teaching, consulting, and supervising research students. His major areas of interest are prediction of human performance and optimization of production systems including occupational safety. He did research for Labour Safety Council of Ontario and taught safety courses in North Atlantic Treaty Organization's (NATO) Summer Institute. He has offered short courses to industry in occupational safety and has developed a personal computer based safety management information system. He has published extensively in his areas of interest and has been invited to present papers at international conferences. He is a member of the editorial boards of leading professional journals in his field of interest.

Dr. B.S. Dhillon is a full Professor and Chairman of Mechanical Engineering and Director of the Engineering Management Program at the University of Ottawa. He has published over 225 articles on Reliability Engineering and related areas. He is or has been on the editorial boards of several journals. In addition, he has written 14 books on various aspects of system reliability, safety, human factors, maintainability and engineering management, published by Wiley (1981), Van Nostrand (1982), Butterworth (1983), Marcel Dekker (1984), Pergamon (1986), etc. His books on Reliability have been translated into several languages including Russian, Chinese, and German. He has served as General Chairman of two international conferences on reliability and quality control held in Los Angeles and Paris in 1987.

Dr. Dhillon is a recipient of the American Society for Quality Control Austin J. Bonis Reliability Award, the Society of Reliability Engineers' Merit Award, the Gold Medal of Honour (American Biographical Institute), and Faculty of Engineering Galinski Award for Excellence in Reliability Engineering Research. He is a registered Professional Engineer in Ontario and is listed in the American Men and Women of Science, Men of Achievements, International Dictionary of Biography, Who's Who in International Intellectuals and Who's Who in Technology, etc.

Dr. Dhillon attended the University of Wales where he received a B.S. in electrical and electronic engineering and a M.S. in mechanical engineering. He received a Ph.D. in industrial engineering from the University of Windsor.

TABLE OF CONTENTS

1   **Introduction** 1
    1.1   History of Safety 1
    1.2   Journals, Books, Organizations, and Data
          Sources Concerned with Safety 3
          1.2.1   Journals 3
          1.2.2   Books 4
          1.2.3   Organizations 6
          1.2.4   Data Sources 7
    1.3   Terms and Definitions 7
    1.4   Scope of the Book.................................... 9
    1.5   Problems 9

2   **Safety Performance Criterion** 11
    2.1   Introduction 11
    2.2   Criterion for Safety Performance.................... 12
          2.2.1   Need for Safety Criterion 12
          2.2.2   Necessary Properties of a Criterion......... 13
    2.3   Measurement and Scales of Measurement............ 13
    2.4   Role of Measurement in Safety Performance 14
    2.5   Desirable Attributes of Measurement
          Techniques 16
    2.6   Commonly Used Measures of Safety 16
          2.6.1   Contemporary Safety Measurement
                  Indices 16
                  2.6.1.1   American National
                            Standards Institute
                            (ANSI) Indices 17
                  2.6.1.2   Bureau of Labor
                            Statistics (BLS)/OSHA
                            Rates 20
                  2.6.1.3   Occupational Injury
                            and Illness Frequency
                            and Severity Rates.............. 21
                  2.6.1.4   Limitations of the
                            Contemporary Methods 23
          2.6.2   Accident Cost Measurements 25
    2.7   Problems 30

3   Statistical Analysis 33
   3.1   Introduction 33
   3.2   Random Variables and Distributions...................34
       3.2.1   Continuous and Discrete Random
           Variables 34
       3.2.2   Sample and Sample Space 35
       3.2.3   Frequency 36
       3.2.4   Probability 36
       3.2.5   Probability Density Function.................37
       3.2.6   Discrete Probability Density
           Functions 37
       3.2.7   Continuous Probability Distribution
           Functions 39
       3.2.8   Characteristics of Random
           Variables 41
           3.2.8.1   Expected Value..................41
           3.2.8.2   Variance 42
   3.3   Selective Probability Distributions 43
       3.3.1   Poisson Distribution 43
       3.3.2   Normal Distribution 44
       3.3.3   Distribution of Sample Means 46
       3.3.4   The Normal Distribution of $\overline{X}$ 47
       3.3.5   The t Distribution............................48
   3.4   Tests of Hypotheses....................................50
       3.4.1   Type I and Type II Errors and
           Steps for Performing a Test of
           Hypothesis 50
       3.4.2   Forms of Hypothesis.........................51
       3.4.3   The Test Statistics 51
       3.4.4   Hypothesis Concerning Two Means........54
   3.5   Control Charts 55
   3.6   Problems 59

4   Safety Behavior Sampling..............................61
   4.1   Introduction 61
   4.2   Fundamentals of Safety Behavior Sampling..........62
   4.3   Procedure for Safety Behavior Sampling.............65
       4.3.1   Define Work Stations.......................65
       4.3.2   Prepare a List of Unsafe Acts 65
       4.3.3   Conduct a Pilot Study 65
       4.3.4   Conducting Additional
           Observations 68

        4.3.5 Correlated Work 69

        4.3.6 Safety Behavior Control Chart..............71

        4.3.7 Improving Safety Behavior..................72

  4.4 Problems 72

5   **Unsafe Conditions and Contributing Factors in**
     Accidents 75

  5.1 Introduction 76

  5.2 Detection of Unsafe Conditions........................76

        5.2.1 Formal Safety Inspections...................76

        5.2.2 Statistical Inference of Safety
              Inspections 78

  5.3 Contributing Causes of Accidents......................80

        5.3.1 Supervisor's Safety Performance 80

        5.3.2 Mental Condition of Workers................81

        5.3.3 Physical Condition of Workers 82

        5.3.4 Evaluation of Contributing
              Conditions for Accidents 82

  5.4 Developing A Composite Score for Safety
     Performance 82

  5.5 Problems 85

6   **Safety Information System**...................................87

  6.1 Introduction 87

  6.2 Safety Information System (SIS)......................89

        6.2.1 Data Base....................................89

        6.2.2 Summary of Investigation....................89

        6.2.3 Statistical Analysis of Accident
              Statistics 90

              6.2.3.1 Univariate Distribution..........91

              6.2.3.2 Univariate Distribution
                       by Department...................92

              6.2.3.3 Bivariate Distribution 92

              6.2.3.4 Statistical Calculation 92

              6.2.3.5 Incidence Rates.................92

              6.2.3.6 Statistical Comparison 93

              6.2.3.7 Data Listing....................93

              6.2.3.8 Univariate Listing 94

              6.2.3.9 Univariate Listing by
                       Department 94

              6.2.3.10 Bivariate Listing................94

6.3 Concluding Remarks 94
6.4 Problems 94

7   Automation and Safety 95
7.1 Introduction 95
7.2 The Effect of Automation on the
     Worker's Performance 95
     7.2.1 Scaling Automation Level.................... 97
     7.2.2 Relationship Between Level of
            Automation and Work
            Requirements 98
7.3 Improving Safety in Automated Plants 98
     7.3.1 Sources of Automation Related
            Hazards 99
            7.3.1.1 Control 99
            7.3.1.2 Presence in the
                     Working Envelop of
                     the Machine..................... 99
            7.3.1.3 Human Error................... 100
            7.3.1.4 Electrical, Hydraulic,
                     and Pneumatic Faults 101
            7.3.1.5 Mechanical Hazards........... 101
7.4 Automated Equipment Safety 101
7.5 Some Suggestions for Improving Safety in
     Automated Plants................................... 102
7.6 Problems 102

Appendix A: Chapter 3 Tables 105
Appendix B: List of Unsafe Acts (Chapter 4)..................... 109
Appendix C: Safety Inspection Check List (Chapter 5) 115
Appendix D: Safety Information System Description and the User
Manual, and Selected Codes from ANSI Z 16.2 (Chapter 6) 165

Index 185

CHAPTER 1

Introduction

1.1 HISTORY OF SAFETY

Safety may be defined as the conservation of human life and its effectiveness, and the prevention of damage to items as per mission needs. Recorded effort concerning the improvement of safety has been going on for almost two thousand years as is evident from *Historia Naturalis* written by Pliny, the Elder (A.D. 23–79). For example, in order to stop the inhalation of toxic substances, he called for the wearing of protective masks by workers in question. During the Middle Ages, George Bauer (1492–1555), father of mineralogy, wrote a 12-volume series on mining and metallurgy. In his writings, he pointed out the problem of mine ventilation and offered several approaches for improving it. Bernadino Ramazzini (1633–1714), father of occupational health and safety, discussed various occupational related diseases and suggested several effective preventive actions. One of the earliest safety devices was the miner's safety lamp invented by Humphrey Davy upon request from the Society for the Prevention of Accidents in Coal Mines.[1] The first volunteer fire department in the colonies was organized by Benjamin Franklin in the year 1736, and in 1752, he laid the foundation of a fire insurance company called the Union Fire Company of Philadelphia. During the early 1800s, Zechariah

1

Allen, proprietor of a cotton factory in Rhode Island, set up an insurance company concerning textile factories and employed a team of inspectors to devise ways and means for improving fire safety at policyholders' factories. Massachusetts Legislature passed the first factory inspection law in 1877. This legislation made fire safety, good housekeeping, machine guarding, and guarding of floor openings and shafts mandatory. In 1885, the American Public Health Association was formed to look after aspects of health, including occupational health and safety. Some of the safety-related acts passed in the U.S. were as follows:[1,2]

1. Pure Food and Drug Act — 1906
2. Walsh-Healy Public Contracts Act — 1936
3. Refrigerator Safety Act — 1956
4. Highway Safety Act — 1966
5. Fair Labor Standards Act — 1938
6. Fire Research and Safety Act — 1968
7. Occupational Safety and Health Act — 1970
8. Consumer Product Safety Act — 1972
9. National Traffic and Motor Vehicle Safety Act — 1966
10. Toxic Substances Control Act — 1976
11. Child Protection and Toy Safety Act — 1969
12. Federal Boat Safety Act — 1971
13. Federal Hazardous Substances Act — 1960

A book on industrial safety by H.W. Heinrich[3] was published in 1931. In this text, he stated that accidents result from unsafe actions and unsafe conditions. Over the years in the U.S., many federal bodies concerned with safety have been formed.[2] Some of those are the National Transportation Safety Board (NTSB), Nuclear Regulatory Commission (NRC), Occupational Safety and Health Administration (OSHA), Federal Aviation Agency (FAA), Food and Drug Administration (FDA), Environmental Protection Agency (EPA), and National Highway Transportation Safety and Administration (NHTSA). It is to be noted that the U.S. Department of Defense has also contributed significantly, especially to the system safety, since the early 1950s.[4] In 1962, a military document entitled "System Safety Engineering for the Development of United States Air Force Ballistic Missiles" was published. Subsequently, the Defense Department published several related documents in the 1960s (e.g., MIL-STD-38130,[5] revised MIL-STD-38130A,[6] MIL-STD-882,[7] System Safety Design Handbook,[8]

MIL-STD-58077,[9] and MIL-STD-23069[10]). Over the years, many authors and researchers have contributed to the subject of safety/system safety. Reference 4 provides an extensive list of selected references.

1.2 JOURNALS, BOOKS, ORGANIZATIONS, AND DATA SOURCES CONCERNED WITH SAFETY

This section presents selective information on areas listed above. This information may provide a useful service to readers directly or indirectly.

1.2.1 Journals

In the English language, there are many journals/newsletters partially or fully covering the subject of safety; some of those are listed below.

1. *Australian Safety News*, National Safety Council of Australia, Melbourne, Australia.

2. *Air Force Safety Journal*, Norton United States Air Force Base, CA.

3. *Fire Prevention*, Fire Protection Association, London, U.K.

4. *International Journal of Aviation Safety*, Capstan Press Limited, Exeter, U.K.

5. *Journal of Safety Research*, National Safety Council, Chicago, IL.

6. *Nuclear Safety*, Nuclear Safety Information Center, Washington, D.C.

7. *Product Safety News*, Institute for Product Safety, Durham, NC.

8. *Safety Management Journal*, National Safety Management Society, Alexandria, VA.

9. *Traffic Safety*, National Safety Council, Chicago, IL.

10. *Accident Analysis and Prevention*, Pergamon Press, Elmsford, NY.

11. *Hazardous Materials Journal*, Elsevier, Amsterdam, The Netherlands.

12. *Hazard Prevention*, System Safety Society, Seabrook, TX.

13. *National Safety News,* National Safety Council, Chicago, IL.

14. *Safety Canada,* Canada Safety Council, Ottawa, Canada.

15. *Accident Prevention,* Industrial Accident Prevention Association, Toronto, Canada.

16. *Family Safety,* National Safety Council, Chicago, IL.

17. *Mine Safety and Health,* U.S. Government Printing Office, Washington, D.C.

18. *Safety Practitioner,* Victor Green Publications, London, U.K.

19. *Accident Facts,* National Safety Council, Chicago, IL.

20. *Safety Surveyor,* Victor Green Publications, London, U.K.

21. *OSHA Report,* Bureau of National Affairs, Washington, D.C.

22. *Occupational Accidents Journal,* Elsevier, Amsterdam, The Netherlands.

23. *Farm Safety Review,* National Safety Council, Chicago, IL.

24. *Occupational Health and Safety,* Medical Publications, Waco, TX.

25. *Journal of Fire Safety,* Technomic, Lancaster, PA.

1.2.2 Books

Since the inception of the safety discipline, many texts concerned directly or indirectly with the field have appeared; selected from those, some are listed below.

1. Heinrich, H.W., *Industrial Accident Prevention,* 4th Ed., McGraw-Hill, New York, 1959.

2. Kuhlman, A., *Introduction to Safety Science,* Springer-Verlag, Berlin, 1989.

3. Asfahl, C.R., *Industrial Safety and Health Management,* Prentice-Hall, Englewood Cliffs, NJ, 1990.

4. Girmaldi, J.V. and Simonds, R.H., *Safety Management,* Richard D. Irwin, Homewood, IL, 1974.

5. Tarrants, W.E., *The Measurement of Safety Performance,* Garland STPM Press, New York, 1980.

6. Dhillon, B.S., *Robot Reliability and Safety,* Springer-Verlag, New York, 1991.

7. Dhillon, B.S., *Reliability Engineering in Systems Design and Operation,* Van Nostrand Reinhold, New York, 1983, chapter 7.

8. Gloss, D.S. and Wardle, M.G., *Introduction to Safety Engineering,* John Wiley & Sons, New York, 1984.

9. Roland, H.E. and Moriarty, B., *System Safety Engineering and Management,* John Wiley & Sons, New York, 1983.

10. Hammer, W., *Product Safety Management and Engineering,* Prentice-Hall, Englewood Cliffs, NJ, 1980.

11. Hammer, W., *Occupational Safety Management and Engineering,* Prentice-Hall, Englewood Cliffs, NJ, 1981.

12. Miller, D.F., *Safety: An Introduction,* Prentice-Hall, Englewood Cliffs, NJ, 1982.

13. Ferry, T.S., *Safety Program Administration for Engineers and Managers,* Charles C. Thomas, Springfield, IL, 1984.

14. Brown, D.B., *Systems Analysis and Design for Safety,* Prentice-Hall, Englewood Cliffs, NJ, 1976.

15. Handley, W., *Industrial Safety Handbook,* McGraw-Hill, London, 1969.

16. Mendelhoff, J., *Regulating Safety,* MIT Press, Cambridge, MA, 1979.

17. Rodgers, W.P., *Introduction to System Safety Engineering,* Wiley, New York, 1971.

18. De Reamer, R., *Modern Safety Practices,* Wiley, New York, 1958.

19. Malasky, S.W., *System Safety: Planning, Engineering, and Management,* Hayden, Rochelle Park, NJ, 1974.

20. Cole, R.A., *Industrial Safety Techniques,* West Publishing Company, Sydney, Australia, 1975.

21. Denton, K., *Safety Management: Improving Performance,* McGraw-Hill, New York, 1982.

22. Wells, G.L., *Safety in Process Plant Design,* Wiley, New York, 1980.

23. ReVelle, J., *Safety Training Methods,* Wiley, New York, 1980.

24. Marshall, G., *Safety Engineering,* Wadsworth, Monterey, CA, 1982.

25. Browning, R.L., *The Loss Rate Concept in Safety Engineering,* Marcel Dekker, New York, 1980.

26. Dhillon, B.S., *Human Reliability with Human Factors,* Pergamon Press, New York, 1986.

1.2.3 Organizations

There are many organizations in various parts of the world concerned with safety; some of these are as follows:

1. World Safety Organization, P.O. Box No. 1, Lalong Laan Building, Pasay City, Metro Manila, The Philippines.

2. National Safety Council, 444 North Michigan Avenue, Chicago, IL.

3. Inter-American Safety Council, 23 Park Place, Englewood, NJ.

4. British Safety Council, 62 Chancellors Road, London W6 9RS, U.K.

5. System Safety Society, 14252 Culver Dr., Suite A-261, Irvine, CA.

6. American Society of Safety Engineers, 850 Busse Highway, Park Ridge, IL.

7. Center for Auto Safety, 1223 Dupont Circle Building, Washington, D.C.

8. Air Safety Division, Airline Pilots Association, 1625 Massachusetts Avenue NW, Washington, D.C.

9. System Effectiveness and Safety Technical Committee, American Institute of Aeronautics and Stronautics, 370 L'Enfant Promenade SW, Washington, D.C.

10. Robotics Institute of America, One SME drive, P.O. Box 930, Dearborn, MI.

11. British Robot Association, 35-39 High Street, Kempston, Bedford, U.K.

1.2.4 Data Sources

This section lists selective data sources directly or indirectly concerned with safety.

1. Loss Management Information System, Gulf Canada Limited, 800 Bay Street, Toronto, Ontario, Canada.

2. Division of Technical Services, National Institute for Occupational Safety and Health, 4676 Columbia Parkway, Cincinnati, OH.

3. Safety Science Abstract Journal, Cambridge Scientific Abstract, Inc., 5161 River Road, Washington, D.C.

4. Nuclear Safety Information Center, Oak Ridge National Laboratory, P.O. Box Y, Oak Ridge, TN.

5. National Technical Information Service, 5265 Port Royal Road, Springfield, VA.

6. Safety Research Information Service, National Safety Council, 444 North Michigan Avenue, Chicago, IL.

7. Computer Accident/Incident Report System, System Safety Development Center, EG&G, P.O. Box 1625, Idaho Falls, ID.

8. International Occupational Safety and Health Information, Center Bureau, International du Travail, CH-1211 Geneva 22, Switzerland.

9. Health and Safety Executive Line, ESA Information Retrieval Service, Via Galileo 00044, Frascati, Rome, Italy.

10. National Electronic Injury Surveillance System, U.S. Consumer Product Safety Commission, 5401 Westbard Street, Washington, D.C.

11. Canadian Center for Occupational Health and Safety, Ministry of Transportation, 400 University Avenue, Toronto, Ontario, Canada.

1.3 Terms and Definitions

There are many terms used in the discipline of safety and their understanding is essential to those concerned with the subject one way or the other. Therefore, this section presents selected terms and definitions related to safety.

Safety: Conservation of human life and its effectiveness, and the prevention of damage to items as per mission needs.

Safeguard: A barrier guard, device, or procedure developed for the purpose of protecting humans.

System Safety: The optimum level of safety subject to resource and operational effectiveness constraints attained through the application of engineering and system safety management principles during the system life cycle.

Hazard: A real or potential situation that may cause unintentional injury or deaths to people or damage to, or loss of, an item or belongings.[11]

Accident: An undesired and unplanned event.[12]

Unsafe Act: The performance of a task subject to less than environment.[1]

Unsafe Condition: Any condition, under the right set of conditions, that will lead to an accident.

Safety Assessment: Concerned with qualitative/quantitative evaluation of safety.

Human Error: The failure to perform a required task (or the performance of a forbidden action) that could result in disruption of scheduled operations or damage to item or property.

Control Chart: The chart that presents control limits.

Sample: A class of items selected randomly and normally from a lot.

Sample Size: The group of items selected randomly from a lot to comprise a single sample.

Standard: Rule(s) or principle(s) applied as a requirement or as a basis for judgement.

Code: A collection of laws, standards, or criteria relating to a specific topic, e.g., safety.

Safety Function: A function carried out by equipment which must work on at least a required lowest level to stop accident occurrences.

Sample Test: A test conducted on a select group of items during the production process.

Control Limit: A limit on a control chart for passing judgement whether the statistical measure, taken from the sample, is inside allowable margins.

Confidence Interval: The upper and lower confidence limits for the containment of a specific value.

Confidence Level: Normally a preassigned probability that the actual value for a population parameter is contained between the given lower and upper limits of the confidence band.

1.4 SCOPE OF THE BOOK

In modern times, increasing attention has been paid to safety in many sectors of industry. As a direct or indirect consequence, hundreds of articles and reports, including over three dozen journals and newsletters in the English language alone, have appeared. However, surprisingly there have been a relatively small number of textbooks published on the subject. Safety is an important concern of almost every organization, as a very large number of people are directly or indirectly concerned with safety in their day-to-day work environments. At present, safety professionals and others needing information on current quantitative safety assessment generally face a great deal of difficulty because they have to study various specialized articles or related documents. This book is an attempt to overcome this difficulty and to present quantitative safety assessment effectively. Generally, previous knowledge is not necessary to digest its contents, since introductory material is provided in several chapters. Topics treated in the book are of current interest. This book should be useful to all people concerned with safety, e.g., professionals, students, management, and academics.

1.5 PROBLEMS

1. Describe in detail the history of the safety discipline.
2. Discuss the following documents:

    A. MIL-STD-882
    B. MIL-STD-58077
    C. MIL-STD-38130

3. Define the following terms:

    A. Safety analysis
    B. Accident analysis
    C. Safety
    D. Safety engineering

4. Write an essay on OSHA.

REFERENCES

1. Gloss, D.S. and Wardle, M.G., *Introduction to Safety Engineering,* John Wiley & Sons, New York, 1984.
2. Hammer, W., *Product Safety Management and Engineering,* Prentice-Hall, Englewood Cliffs, NJ, 1980.
3. Heinrich, H.W., *Industrial Accident Prevention,* 4th ed., McGraw-Hill, New York, 1959.
4. Dhillon, B.S., *Reliability Engineering in Systems Design and Operation,* Van Nostrand Reinhold Company, New York, 1983.
5. MIL-STD-38130, Safety Engineering of Systems and Associated Sub-Systems and Equipment-General Requirements (proposed), U.S. Department of Defense, Washington, D.C., Sept. 30, 1963.
6. MIL-STD-38130A, Safety Engineering of Systems and Associated Subsystems and Equipment-General Requirements, U.S. Department of Defense, Washington, D.C., March 1966.
7. MIL-STD-882, Systems Safety Program for System and Associated Subsystem and Equipment-Requirements, July 15, 1969.
8. AFSC DH1-6 AFSC, Design Handbook-System Safety, U.S. Department of Defense, Washington, D.C., July 1967.
9. MIL-STD-58077, Safety Engineering of Aircraft System, Association Subsystem and Equipment-General Requirements, U.S. Department of Defense, Washington, D.C., October 26, 1967.
10. MIL-STD-23069, Safety Requirements Minimum for Air Launched Guided Missiles, U.S. Department of Defense, Washington, D.C., October 31, 1965.
11. Omdahl, T.P., *Reliability, Availability, and Maintainability (RAM) Dictionary,* American Society for Quality Control (ASQC) Quality Press, Milwaukee, WI, 1988.
12. Roland, H.E. and Moriarty, B., *System Safety Engineering and Management,* Wiley, New York, 1983.

CHAPTER 2

Safety Performance Criterion

2.1 INTRODUCTION

Safety is considered to be a commonsense approach to removing agents of injury. Safety, as a concept and practice, has shifted to a complex methodology for the reliable control of injury to human beings and damage to property. However, it does lack a theoretical base — this may be due to the fact that this concept is still going through the transitional phase[1] just like other fields of sciences of the past.

As safety is concerned with reducing accidents and controlling or eliminating hazards at the work place, accident prevention is a significant step toward safety improvement. Nowadays, there is tremendous pressure exerted on companies to improve safety. Major factors responsible for this pressure are economical, social, and government regulations. An understanding of accident causation is a prerequisite for safety improvement.

Many attempts have been made over the years to develop a predictive theory of accident causation. Researchers addressing the accident phenomena have come up with conflicting theories. Hale and Hale[2] and Surrey[3] have conducted extensive reviews of the pertinent literature. By and large, most theories proposed are based on two distinct models:

behavioral and situational. Behavior models consider humans as the major, if not the sole, factor responsible for the causation of an accident. On the other hand, the situational models consider the interactions between humans, environment, and situation for studying the accident process. The number of variables that influence any given accident situation are too numerous, and their interaction cannot be quantified due to lack of knowledge required. Once the individual components involved in an accident causation are better understood and represented in quantitative terms, it may be possible to quantify the relationships involved in the overall process of accident causation. Until then, all the theories of accident causation will still be conceptual in nature and thus will remain of limited use in preventing and controlling accidents.

Measurement is an essential prerequisite for controlling and preventing accidents. Tarrants[4,5] describes current attempts to control accidents on "trial and error" because of lack of adequate control measures of the effectiveness. Other related literature may be found in References 6 and 7.

2.2   CRITERION FOR SAFETY PERFORMANCE

Two important factors considered here are the need for safety criterion and its associated necessary properties. Both of these items are separately discussed below.

2.2.1   Need for Safety Criterion

Safety is considered to be a measure of relative freedom from accidents and to improve safety performance; control of accidents is essential. For checking the effectiveness of control, the measurement is important and "What to measure?" is commonly termed as criterion.

There are many reasons for having the safety performance criterion. Some of these are:

1. carrying out comparisons
2. estimating forecasts
3. conducting trend analysis
4. evaluating safety improvement program effectiveness
5. identifying problem areas
6. optimal allocation of resources for improving safety performance

2.2.2 Necessary Properties of a Criterion

The two major properties that a criterion must possess are reliability and validity.

A criterion is considered to be adequately reliable if, for an error-free measurement technique, it is capable of duplication with the same results obtained from successive applications to the same situation. If the reliability of an individual safety criterion is satisfactory, this means that workers can be graded in terms of accident susceptibility without considering their exposure experience and accident-related characteristics.

A criterion is said to be valid if, by its very nature, it is satisfactorily suited to the object being measured. Sanders and McCormick[8] define validity as "the extent to which the measure in question is considered to be a relevant or pertinent index of criterion in mind, such as, system performance, quality of work, comfort in seating, or job satisfaction".

2.3 MEASUREMENT AND SCALES OF MEASUREMENT

Measurement may be defined as any process which involves the assignment of numerals to objects or events according to rules.[9] The essential function of measurement is the setting in order of a class of events with respect to their exhibition of a particular property, and this entails the discovery of an ordered class the elements of which can be put in a one-to-one correlation with the events in question.

For the purpose of safety performance, the measurement should enable one to compare the same accident producing characteristics of different objects, the accident producing characteristics of the same object at different times, and to describe how accident-producing characteristics of the same or different things are related to each other.

Now, suppose one is considering studying two situations from the safety point of view (assuming that accurate techniques for measuring safety characteristics of such situations are available). In this case, if interest is in only determining "equality" of the two situations, then the scale used will be termed as ordinal. On the other hand, if the interest is to determine the equality of intervals or differences, the scale used will be known as "interval". Furthermore, if the safety of one situation is twice that of the other, then the scale used will be known as the ratio scale. This scale embraces all the characteristics of interval and ordinal scales. When the measurements conform to the requirements of the ratio scale, then one should note here that all types

Table 2.1 Scales of Measurement Categories

Scale	Basic empirical operation	Examples
Nominal	Equality determination	"Numbering" of departments
Ordinal	Determination of greater or less	Test scores
Interval	Determination of the equality of intervals or of differences	Temperature, intelligence test "standard scores"
Ratio	Determination of the equality of ratios	Length, work, density, intervals, time, etc.

For more details, consult Stevens, S.S., Measurement, psychophysics, and utility, in *Measurement Definitions and Theories,* Churchman, C.W. and Ratoosh, A., Eds., John Wiley & Sons, New York, 1959.

of statistical measures are only applicable. A classification of scales as developed by Stevens[9] is given in Table 2.1.

2.4 ROLE OF MEASUREMENT IN SAFETY PERFORMANCE

Safety performance is related to accident prevention. Before we discuss accident prevention, let us examine the necessary ingredients of an accident as well as some of the characteristics of the ingredients involved.

For an accident to take place, the following must be present:

1. worker
2. machine, tools, or equipment
3. physical environment
4. social environment

If A_L denotes the Accident Level indicating the degree of losses, i.e., property, injury, etc., then:

$$A_L = \int (W_{ij}), (O_{tk}), (P_{el}), (S_{gm})$$

Each of the above four factors are described below where:

W_{ij} is the combined effect of ith worker's characteristics "j" as related to safety. Some examples of the worker characteristics are physical, abilities and skills, interests, and personality traits. Physical characteristics include anthropometrical dimensions, visual acuity, hearing, and so on. Such characteristics can be measured with a high degree of accuracy. Abilities and skills include dexterity, verbal ability, intelligence, mechanical aptitude, etc. and can be measured within a medium degree of accuracy. Interests and personality traits include scientific interests, safety attitude, tolerance, and emotional stability; like characteristics such as these can only be measured with a low degree of accuracy.

O_{tk} is the combined effect of the Oth machine's characteristic "k" as related to safety. These characteristics include mechanical actions, location of machine controls, quality of maintenance, and other related items. Characteristics such as these may be so designed that for a given set of conditions, accident prevention is maximized. This category also includes safety-related characteristics of "materials" and energy used.

P_{el} is the combined effect of physical environment "e" at location "l" as related to safety. Some of the components of these characteristics are temperature, humidity, illumination, air contamination, and the state of housekeeping. Individual characteristics may be measured in ratio scales and may be taken as indirect measures/indicators of the safety performance.

S_{gm} is the combined effect of the social environment of characteristics "g" at location "m" as related to safety. Regulations, formal rules, and laws that influence worker behavior are the typical examples of such characteristics.

As all types of measurements are subject to certain errors, there are many possible causes[10] of such errors: the observer, the measuring instrument, the environment, and the object being observed, and so on.

Accuracy is the measure of the extent to which a given measurement deviates from what really is being measured. Based upon such measurements, predictions are made and the accuracy of such predictions may be defined by means of a confidence interval. This interval helps reveal that a specific range of numbers based on measurements has a certain probability of including the measurement.

From an accident point of view, the relationship(s) among W_{ij}, O_{tk}, P_{el}, and S_{gm} is largely dependent upon the existence of various measures. These measurements are needed to evaluate the safety performance in any organization.

2.5   DESIRABLE ATTRIBUTES OF MEASUREMENT TECHNIQUES

When developing an appropriate measurement technique, it is essential to pay careful attention to defining the desired performance. Tarrants[4] has postulated a number of characteristics of a good measurement technique as follows:

1. administrative feasibility

2. adaptability in the range of characteristics to be evaluated

3. constant unit of measurement throughout the range to be evaluated

4. quantifiable measurement criterion

5. sensitive measurement technique

6. capability of duplication with the same results obtained from the same items measured

7. inclusion of the validity of the measurement

8. error-free results

9. efficient and understandable

The above nine characteristics demonstrate that the solution to the problem of finding an effective measurement technique is quite complex. In any case, it is quite unlikely to find all the desired characteristics in a single measurement device, but every effort should be directed towards having a technique with all such possible characteristics.

2.6   COMMONLY USED MEASURES OF SAFETY

This section presents selective and commonly used measurement techniques. It appears that none are perfect, but some are better than others under given conditions.

2.6.1   Contemporary Safety Measurement Indices

This section discusses various types of indices: American National Standards Institute document ANSI Z 16.1, Bureau of Labor Statistics

Table 2.2 Injuries and Illnesses Experienced by Workers

Type of injury or illness	No. of injuries or illnesses	Days lost or charged
Fractures	5	75
Ankle twists	10	150
Thumb amputations at the distal phalanx	4	1200
Cases of dermatitis	7	50
Total	26	1475

(BLS)/Occupational Safety and Health Act (OSHA), and Occupational Injury and Illness. All these three types of indices are discussed below.

2.6.1.1 American National Standards Institute (ANSI) Indices

Prior to the advent of the OSHA, the ANSI Z 16.1-1967 (R 1973) Standard for Recording and Measuring Injuries was developed by the American National Standards Institute. In 1970, OSHA called for mandatory record-keeping requirements very similar to the ones given in ANSI Z 16.1.[11]

The ANSI system uses frequency and severity rates that pertain to death, disabling (lost time) injuries (including total, permanent partial, temporary total, and temporary partial). The rates are based on a schedule of charges for deaths, permanent total, permanent partial disabilities, plus the total days of disability for all temporary total disabilities. These measures pertain to the relative frequency of occurrence of major injuries and days lost or charged to them. The Disabling-Injury Frequency Rate (DIFR), the Disabling Injury Severity Rate (DISR), and Average Days Charged (ADC) are defined below.

DIFR is defined as the number of disabling injuries (including illness) per million employee hours worked:

$$\text{DIFR} = \frac{(\# \text{ of disabling injuries}) \text{ (one million)}}{(\# \text{ of employee hours worked})} \tag{2.1}$$

Example 2.1

Suppose in a given year, XYZ company employed 500 full-time workers and 200 half-time workers. Table 2.2 is the record of injuries and illnesses experienced by the workers during that year.

Using the data given in Table 2.2, we get

# of disabling injuries = 26

# of employee hours worked =
(# of full time employees) (40 hours/week) (50 weeks/year) +
(# of 1/2 time employees) (40 hours/week) (25 weeks/year)
= (500) (40) (50) + (200 (40) (25)
= 1,200,000 hours

Using the above calculations in Equation 2.1 yields

$$\text{DIFR} = \frac{(26)\ (10^6)}{1,200,000} = 21.67$$

For accidents resulting in deaths, permanent total, and permanent partial disabilities, the days charged are determined by the American Standards Scale of Time Charges. A schedule of charges for loss of member — traumatic or surgical — and impairment of function is provided in Table 2.3.

DISR is defined as the number of days lost or charged per million employee hours worked. Days lost include all scheduled charges for all deaths, permanent total, and permanent partial disabilities, plus the total days of disability from all temporary total injuries which occur during the period covered. Thus, we have

$$\text{DISR} = \frac{(\text{total days charged})\ (10^6)}{(\text{\# employee hours})} \tag{2.2}$$

Example 2.2
Calculate DISR for the data given in Example 2.1 for XYZ company. This is computed as follows:

# of days lost or charged = 1475
# of employee hours worked = 1,200,000

Thus, we get

$$\text{DSIR} = \frac{(1475)\ (10^6)}{1,200,000} = 1229.16$$

For XYZ company, it is apparent that for that particular year it had 21.67 disabling injuries and 1229.16 days lost or charged for every million man-hours worked.

Table 2.3 Tabulation of Scheduled Charges in Days

A. For loss of member — traumatic or surgical

A.1 Fingers, Thumb, and Hand (amputation involving all or part of the bone)

Description (amputation involving all or part of bone)	Hand	Thumb	Fingers			
			Index	Middle	Ring	Little
Distal phalange	—	300	100	75	60	50
Middle phalange	—	—	200	150	120	100
Proximal phalange	—	600	400	300	240	200
Metacarpal	—	900	400	300	240	400
Hand at wrist	3000					

A.2 Toe, Foot, and Ankle (amputation involving all or part of the bone)

	Great toe	Each of other toes
Distal phalange	150	35
Middle phalange	—	75
Proximal phalange	300	150
Metatarsal	600	350
Foot at ankle, 2400		

A.3 Arm

Any point above + elbow, including shoulder joint	4500
Any point above wrist and at or below elbow	3600

A.4 Leg

Any point above + knee	4500
Any point above ankle and at or below knee	3000

B. Impairment of Function

Description	
One eye (loss of sight), whether or not there is sight in the other eye	1800
Both eyes (loss of sight), in one accident	6000
One ear (complete industrial loss of hearing), whether or not there is hearing in the other ear	600
Both ears (complete industrial loss of hearing), in one accident	3000
Unrepaired hernia (for repaired hernia, use actual days lost)	50

ADC can be computed to indicate the average length of disability per disabling injury as follows:

$$\text{ADC} = \frac{\text{\# of days lost or charged}}{\text{\# of disabling injuries}}$$

$$= \frac{\text{DISR}}{\text{DIFR}} \qquad (2.3)$$

Example 2.3

Compute the value of ADC using the results of Examples 2.1 and 2.2. Thus, we get

$$ADC = \frac{1229.16}{21.76} = 56.49$$

2.6.1.2   Bureau of Labor Statistics (BLS)/OSHA Rates

With the passage of OSHA, a new method of measurement has come into existence. The BLS-OSHA system of measuring safety and health performance is intended to serve as a nationwide survey of work injuries and illnesses and to provide national statistics on an industry basis for all recordable occupational injuries and illnesses occuring at the work place. Any occupational injuries and illnesses that result in death, regardless of time between the injury and death or the length of illness, nonfatal occupational illnesses other than fatalities that result in lost work days, and occupational injuries that result in transfer to another job or require medical treatment are considered recordable cases. A guide to recordability of cases under OSHA is shown in Figure 2.1.   The OSHA incident rate, IR_{OSHA}, is defined by

$$IR_{OSHA} = \frac{\# \text{ of recordable injuries} \times (200,000)}{\# \text{ of employee hours worked}} \tag{2.4}$$

It is to be noted that in the above formula, the number 200,000 is the equivalent to 100 full-time employees at 40 hours per week for 50 weeks.

Example 2.4

Suppose that a company employing 350 full-time employees experienced the accidents and illnesses given in Table 2.4 during 1986. Calculate incident rate using Equation 2.4.

Using the above data in Equation 2.4, we get the following value for the incident rate:

$$IR_{OSHA} = \frac{(13)\,(200,000)}{(350)\,(40)\,(50)} = 3.71$$

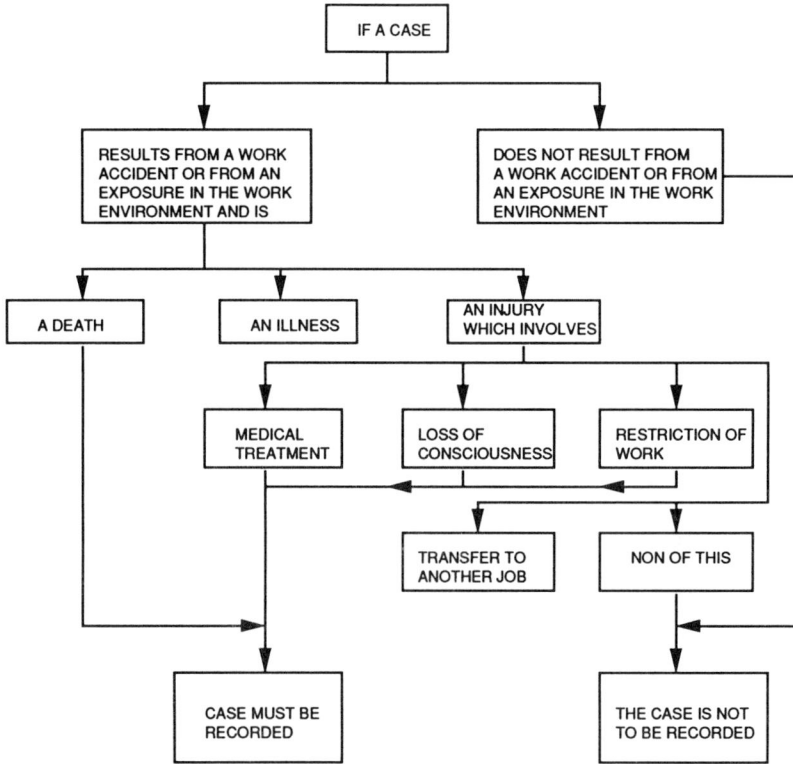

Figure 2.1. Guide to recordability of cases under OSHA.

Table 2.4 Accidents and Illnesses

Injury/illness	Number	Days lost
Broken ankle	1	25
Carbon monoxide poisoning	1	2
Dermatitis cases	5	0
Welding flash burns	2	4
Cuts requiring stitches	4	0
Total	13	31

2.6.1.3 Occupational Injury and Illness Frequency and Severity Rates

Both these rates are defined as follows:

$$\text{Frequency Rate} = \frac{(\text{\# of lost time injuries}) \ (200{,}000)}{(\text{\# of employee hours worked})} \quad (2.5)$$

Table 2.5 Data for the Company in Example 2.5

Division 1 Description	Days lost	No. of injuries/ illness	Division 2 Description	Days lost	No. of injuries/ illness
Sprained ankle	1	1	Broken ankle	10	1
Broken thumb	2	1			
Eye injury due to embedded metal	0	1			
Cuts requiring stitches	3	4			
Loss of consciousness	0	2			
Cases of dermatitis	4	4			
Total	10	13		10	1

and

$$\text{Severity rate} \quad = \frac{(\# \text{ of days lost}) \, (200{,}000)}{(\# \text{ of employee hours worked})} \quad (2.6)$$

Example 2.5

Suppose that a company has two divisions, each employing 100 employees. In 1989, Division A had 15 lost-time injuries and Division B had 3 deaths. Calculate values of frequency rates using Equation 2.5.

Using the above data in Equation 2.5 we get the following values for frequency rates (FR):

$$FR = \frac{15 \times 200{,}000}{200{,}000} = 15 \qquad FR = \frac{3 \times 200{,}000}{200{,}000} = 3$$

Example 2.6

Suppose for the company in Example 2.5, we have the data given in Table 2.5. Using the data of Table 2.5 in Equation 2.6 we get the following results for severity rate (SR):

$$SR = \frac{10 \times 200{,}000}{200{,}000} = 10 \text{ and}$$

$$SR = \frac{10 \times 200{,}000}{200{,}000} = 10$$

Examples 2.5 and 2.6 illustrate the fact that rates and their comparisons may be deceiving.

Note that the following incidence rates are also used by some organizations:

1. injury incidence rate
2. illness incidence rate
3. fatality incidence rate
4. lost work-days cases incidence rate
5. number of lost work-days rate
6. specific hazard incidence rate

All these rates use the standard 200,000 factor.

2.6.1.4 Limitations of the Contemporary Methods

Some of the limitations[5] of the existing measures are as follows:

1. The rates in question are not sensitive enough to serve as an accurate indicator of safety effectiveness.

2. The smaller the work force, the less reliable are the rates as an indicator of safety performance.

3. Lost-time accidents, deaths, and other recordable injuries are events of relatively rare occurrence.

4. A single severe injury or death will drastically alter the severity rate in smaller organizations, and this rate may not accurately reflect overall accident prevention accomplishments.

5. The rates do not reflect environments involving nonparallel hazard categories.

6. The rates are based on after-the-fact appraisals of injury-producing accidents.

It may be said that the rates discussed here are not measures of safety performance, but are failure measurements of events that were recorded by chance to meet the legal requirements or by good intentions.

Girmaldi[12] has suggested a three-dimensional appraisal of a company's safety performance using DIFR, DISR, and ADC, thus circumventing the disadvantage of employing only an incident rate. This appears to be a workable solution, to some extent, to a very complex problem of developing a "safety performance measure".

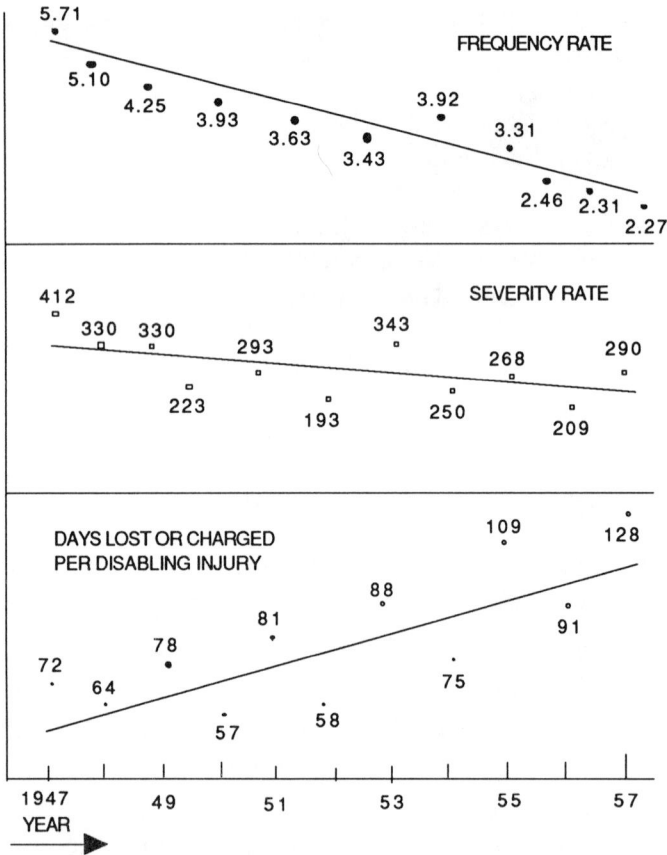

Figure 2.2. Disabling injury trends for period 1947 to 1957 based on data from Girmaldi, J. V., *Safety and Accident Prevention in Chemical Operations,* Fawcett, H. H. and Wood, W. S., Eds., Interscience Publishers, New York, 1964.

Figures 2.2 and 2.3 are based on Girmaldi's[12] data. Figure 2.2 demonstrates that although DIFR and DISR trends are indicating a desirable trend, ADC is increasing. This could be taken as an indicator of having an "injury control program" in place rather than removing all the "hazards". In Figure 2.3, data for the period 1953 to 1963 is presented (the hazard control program was initiated in 1953). In this case, all the measures (i.e., DFIR, DISR, and ADC) have a downward trend — a very desirable situation.

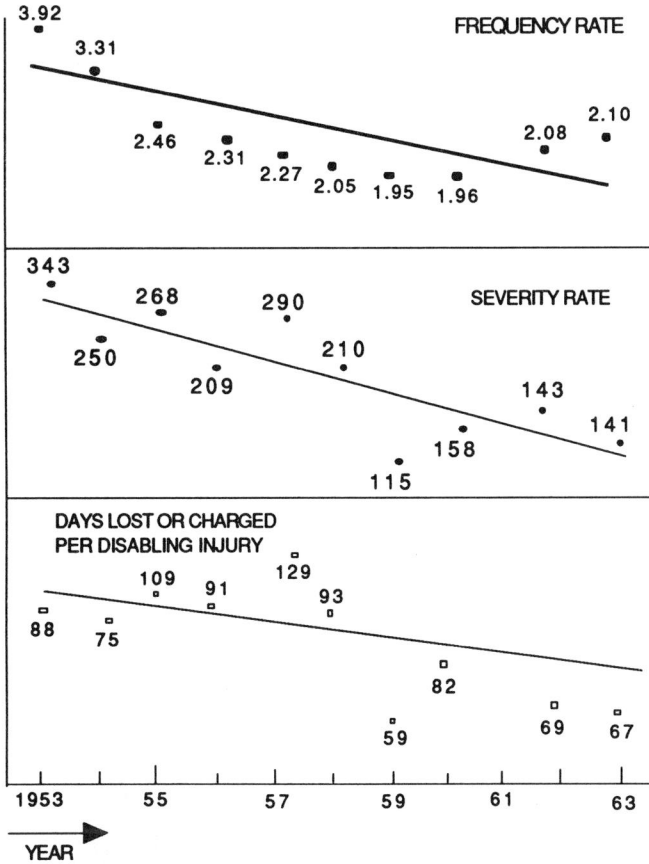

Figure 2.3. Disabling injury trends for period 1953 to 1963 based on data from Girmaldi, J. V., *Safety and Accident Prevention in Chemical Operations,* Fawcett, H. H. and Woods, W. S., Eds., Interscience Publishers, New York, 1964.

2.6.2 Accident Cost Measurements

Generally, the occurrence of accidents leads to numerous losses. In estimating an accident's cost, it may not be practical to determine accurately the cost of these losses, but even approximate total figures will indicate that such costs could be financially damaging. Some of these costs could be accident-investigation costs, payments for settlement of lawsuits, costs of corrective measures to stop recurrences, payments for property damage on top of coverage by insurance, insurance increment costs, legal costs associated with defense against claims, and the costs associated with loss of public confidence (thus, reduction in revenue). Therefore, it is necessary to be aware of the costs of accidents because this awareness helps in justifying the associated expenditures, manpower, etc. Generally, management always aims to reduce this cost to a minimum.

There are several methods of estimating the cost of accidents. Two methods of assessing the costs of an accident are described below.

The Heinrich Method

Over 50 years ago, Heinrich[13] stated that for every dollar of insured cost paid for accidents there were four dollars of uninsured costs borne by the organization. Indirect cost elements, as enumerated by Heinrich, are given below.

1. cost of lost time of injured employee
2. cost of lost time of employees who stop work or are involved in the action
3. cost of lost time by management
4. cost of lost time on the case by first aid and hospital people not paid by insurance
5. cost due to machine/material damage
6. cost due to lost orders, etc.
7. cost to employees under welfare and benefit system
8. cost to employees in continuing wages of the insured
9. cost due to weakened morale
10. cost due to loss of profit and employee productivity
11. overhead cost for injured employee while in nonproduction status

The readers may observe that most of the above costs are open to speculation and, thus, their very existence may be questionable.

Girmaldi and Simonds Method

The insurance costs can easily be ascertained, but the uninsured costs must be estimated by using various ways and means. Girmaldi and Simonds[14] suggested the following formula for estimating the uninsured costs:

Uninsured Cost = A [# of lost work-day cases with days
away from work (lost days)]

+

B [# of doctor's cases (OSHA non-lost workday
cases that are attended by a doctor)]

(2.7)

+

C [# of first aid cases]

+

D [# of non-injury "accidents"]

where A, B, C, and D are constants representing the average cost for each case, respectively. The remaining information expressed in Equation 2.7 is discussed below.

1. Lost time cases: include permanent partial disabilities and temporary total disabilities
2. Doctor's cases: include temporary partial disabilities and medical treatment cases requiring the attention of a physician
3. First aid cases: include medical treatment cases requiring only first aid and those resulting in property damage of less than $20.00 and in loss of less than 8 h of working time
4. No-Injury Cases: include unintended occurrences: (a) resulting in loss of 8 or more man-hours or $30.00 or more property damage, (b) affording danger of personal injury, and (c) by chance not actually causing personal injury or else resulting in minor injury not requiring medical attention.

It may be observed that fatalities and permanent total disabilities are not included in this method for the reason that deaths and permanent disabilities are sufficiently rare in any organization. However, if necessary, such costs may subsequently be added into the calculation.

For developing the values of constants A, B, C, and D, a pilot study may very well be undertaken in an organization. In a situation where the cost for conducting such a study is considered to be high or to outweigh the benefits derived from it, the figures suggested in Reference 14 and given in Table 2.6 may be used.

Table 2.6 Suggested Values for Constants A, B, C, and D

Constant	Suggested value ($)
A	220
B	55
C	12
D	400

Table 2.7 Accident Cost-Related Data for an Organization

General information	Dollars	Accident-related occurrences	Frequency
Average production worker wage (hourly)	20	—	—
Insurance premium	150,000	—	—
Insurance refund	27,000	—	—
—	—	Lost time cases	25
—	—	Doctor's cases	75
—	—	First-aid cases	220
—	—	No injury cases	30

The costs given in Table 2.6 were based on wage and price levels as of February 1974 when average hourly wage for production workers in manufacturing was $4.04. By dividing $4.04 by the current average wage, a multiplier may be derived to adjust each of these costs.

Example 2.7

Suppose that the data given in Table 2.7 were applicable to an organization in 1986. Calculate the accident cost using these data and the Girmaldi and Simonds method.

The total accident cost (AC) is given by

$$AC = \text{Insured Cost (IC)} + \text{Uninsured Cost (UC)} \quad (2.8)$$

The insured cost is defined by

$$IC = \text{Insurance Premium} - \text{Insurance Refund} \quad (2.9)$$

Substituting the given data into Equation 2.9 we get

$$IC = \$150,000 - 27,000 = 123,000$$

Similarly, from Equation 2.7 and Table 2.6 the uninsured cost is

$$UC = A(25) + B(75) + C(220) + D(30) \quad (2.10)$$

Thus, from the given information, the value of the wage adjustment multiplier (WAM) is

$$\text{WAM} = \frac{\$20}{\$4.04} = 4.95$$

The following adjusted values for A, B, C, and D are obtained:

$$A = \$4.95 \times \$220 = \$1089$$
$$B = \$4.95 \times \$55 = \$\ 272.25$$
$$C = \$4.95 \times \$12 = \$\ \ 59.40$$
$$D = \$4.95 \times \$400 = \$1980$$

Using the above values in Equation 2.10, we get

$$UC = 1089\ (25) + 272.25\ (75) + 59.40\ (220) + 1980\ (30)$$
$$= 27,225 + 20,418.75 + 13,068 + 59,400$$
$$= 120,111.75$$

The resulting total accident cost (AC) is

$$AC = IC + UC$$
$$= 123000 + 120111.75$$
$$= \$243112.75$$

It may be possible to determine accident costs more accurately without using the Girmaldi and Simonds method. However, it is still a challenge to determine the optimal accuracy level when the costs associated with obtaining the accuracy required are at a minimum and the benefits derived from that accuracy level are at their maximum.

A valid estimate of accident cost is necessary for developing an effective accident prevention program. All in all, the approach suggested by Girmaldi and Simonds appears to be workable and quite useful.

Optimal Cost-Benefit Model

This approach aims at developing a cost-effective accident prevention program which maximizes the injury and illness reduction within

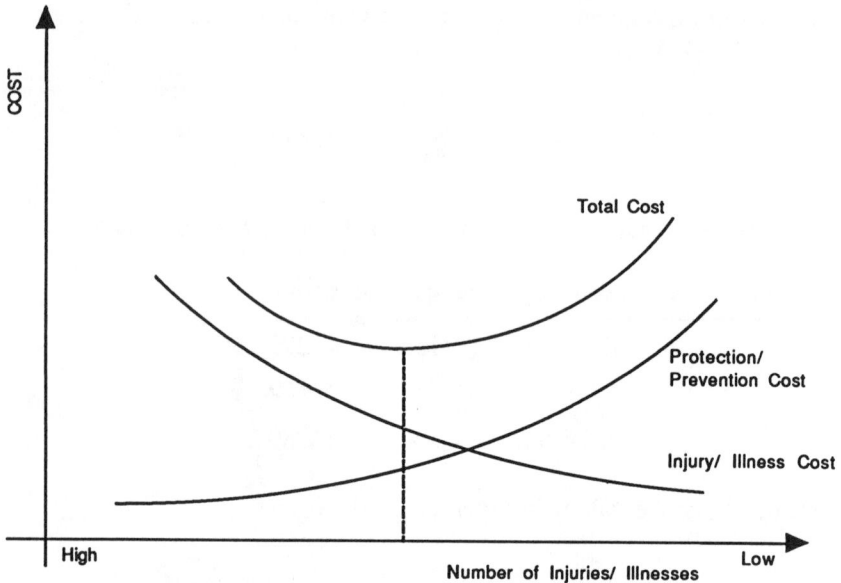

Figure 2.4.   Injuries/illnesses vs. cost.

constraints of available funds. It may be argued here that as more funds are expended for an accident prevention program such as OSHA, the costs of injuries and illness should decrease. This expected relationship is shown in Figure 2.4.

The intersection of the two curves shown in Figure 2.4 denotes the minimum total cost when the most benefit, in terms of reduced injury and illness costs, can be achieved for the minimum investment.

The validity of this approach may be questioned when occupational illnesses take a decade or more to appear. Regulatory bodies (like OSHA) take maximum precautions, and the industry favors the regulatory body to measure these costs and the benefits to workers concerned and to ensure that both are reasonably related.

2.7 PROBLEMS

1. What are needs for having a safety performance criterion?
2. Discuss important properties of the safety performance criterion.
3. List at least nine characteristics of a good measurement (safety) method.

Table 2.8 Accident Cost-Related Data for an Organization for 1991

No.	General information	Dollars	Accident-related occurrences	Frequency
1	Average production work wage (hourly)	20	—	—
2	Insurance refund	50,000	—	—
3	Insurance premium	200,000	—	—
4	—	—	Lost time cases	30
5	—	—	Doctor's cases	80
6	—	—	First-aid cases	230
7	—	—	No injury cases	35

4. Discuss the following:
   A. Disabling-injury frequency rate
   B. Disabling-injury severity rate
   C. Average days charged
5. Compare the OSHA rates with the ANSI document (ANSI Z 161) indices.
6. Describe the following two accident cost estimation methods:
   A. The Heinrich method
   B. The Girmaldi and Simonds method
7. Table 2.8 presents certain data for an organization for 1991. Compute the total accident cost using these data and the Girmaldi and Simonds method.

REFERENCES

1. Kuhlmann, A., *Introduction to Safety Science,* Springer-Verlag, Berlin, 1989.
2. Hale, A.R. and Hale, M., *A Review of Acident Research Literature,* Her Majesty's Stationery Office, London, 1972.
3. Surrey, J., *Industrial Accident Research: A Human Engineering Approach,* University of Toronto, 1968.
4. Tarrants, W.E., *The Measurement of Safety Performance,* Garland STPM Press, New York, 1980.
5. Tarrants, W.E., Applying measurement concepts to the appraisal of safety performance, *J. Am. Soc. Saf. Eng. (ASSE),* 1965.
6. Rockwell, T.H., Problems in the measurement of safety performance, paper presented at the Symp. Measuring Industrial Safety Performance, Ohio State University, Columbus, 1968.
7. Rockwell, T.H. and Bhise, V.P., Two approaches to a non-accident measure for continuous assessment of safety performance, in *The Measurement of Safety Performance,* Tarrants, W.E., Ed., Garland STPM Press, New York, 1970.

8. Sanders, M.S. and McCormick, E.J., *Human Factors in Engineering and Design,* McGraw-Hill, New York, 1990.

9. Stevens, S.S., Measurements, psychophysics and utility, in *Measurement Definitions and Theories,* Churchman, C.W. and Ratoosh, A., Eds., Wiley, New York, 1959.

10. Ackoff, R.C., Gupta, S.K., and Minhas, J.S., *Scientific Method Optimizing Applied Research Decisions,* Wiley, New York, 1962.

11. Asfahl, C.R., *Industrial Safety and Health Management,* Prentice Hall, Englewood Cliffs, NJ, 1990.

12. Girmaldi, J. V., Measuring safety effectiveness, in *Safety and Accident Prevention in Chemical Operations,* Fawcett, H.H. and Wood, W.S., Eds., Interscience Publishers, New York, 1964.

13. Heinrich, H.W., *Industrial Accident Prevention: A Scientific Approach,* McGraw-Hill, New York, 1959.

14. Girmaldi, J.V. and Simonds, R.H., *Safety Management,* Richard D. Irwin, Homewood, IL, 1984.

CHAPTER 3

Statistical Analysis

3.1 INTRODUCTION

Statistics can reduce raw data to manageable forms and allow the study and analysis of variance, thus allowing managers to maximize the use of information available in arriving at an effective decision. Similarly, safety professionals' ability to measure, predict, and control potential factors for minimizing losses in an organization is enhanced by performing statistical analysis of the raw safety-related data.

This chapter presents some of the basic techniques used for performing statistical analysis. Some of these are relevant probability density functions, tests of hypotheses, and control charts. The main objective of this chapter is to demonstrate how the existing statistical analyses techniques can be applied to the safety-related problems. For further reading on statistical analyses techniques, the readers are directed to any standard text on probability and statistics or to documents (References 1 through 6) such as those listed at the end of the chapter.

3.2   RANDOM VARIABLES AND DISTRIBUTIONS

The amount of rainfall in a given year, the demand for a certain product, the life of a light bulb, the quality of one product produced by a certain process, and the number of accidents in a given year for a particular organization are all examples of random phenomena. They are called random because they occur unpredictably. Similarly, the diameters of disks stamped out of sheet metal could vary, for example, from 5.98 to 6.02 in. These variations could be due to many factors, i.e., variations in voltage driving the electric motor running the punch press, variations in the sheet metal thickness, and so on. Whatever the reason, such like phenomena are also known as random.

A variable corresponding to one or several measurable characteristics of a random phenomena is called a random variable. Any particular characteristic measurement, when repeated, is subject to a certain amount of variation. For example, the diameter of the disk stamped out of sheet metal can be identified as a random variable whose value varies from 5.98 to 6.02 in. Similarly, safety measurements in terms of number of lost work-days for a given type of accident will have considerable variations. At a particular stage, let us assume that a safety awareness program is started by the management. After the initiation of such a program, it is likely that the number of lost days for that particular type of accident decreases. This decrease may either be due to random variations or due to the influence of the "awareness program". We need procedures for comparing measurements to determine if the differences between measurements are due to causal effects (in our case, awareness program) or merely to random variations; this is when statistical analysis is useful.

3.2.1   Continuous and Discrete Random Variables

If a random variable can take on any value on a continuum, it is known as a continuous random variable. However, if it takes distinct values only, it is called a discrete random variable. In our earlier example of stamping a disk, the diameter can have only the value between 5.98 and 6.02 in. Thus, a random variable corresponding to this is a continuous one. However, the number of accidents in a given month can be 0, 1, 2, etc., but never 2.8 or 1.9. Thus, the corresponding random variable in this case is discrete.

Table 3.1 Accident Record for Three Years[a]

Time period (month)	No. of accidents
1	1
2	7
3	3
4	4
5	6
6	3
7	5
8	5
9	7
10	6
11	4
12	7
13	6
14	8
15	5
16	2
17	7
18	11
19	6
20	0
21	4
22	7
23	6
24	5
25	3
26	6
27	8
28	9
29	6
30	9
31	4
32	7
33	5
34	6
35	3
36	8

[a] Number of accidents per month.

3.2.2 Sample and Sample Space

Statistical properties of a random phenomenon are usually estimated by means of sampling. Sampling is the process of observing a phenomenon for a finite number of times. A set consisting of all possible values that a random variable can take on is called the sample space of that random variable.

A subset of a sample space is called an event. Table 3.1 shows the accident record for 36 months of a certain company. From this table, we can observe the following:

Table 3.2 Distribution of Monthly Accidents

No. of accidents	Frequency	Probability
0	1	1/36
1	1	1/36
2	1	1/36
3	3	3/36
4	4	4/36
5	5	5/36
6	8	8/36
7	6	6/36
8	3	3/36
9	1	1/36
10	2	2/36
11	1	1/36

1. The number of accidents per month vary from 0 to 11.
2. The table consists of a "sample" of 36 observations.
3. The number of months having no accidents is 1. (This is an event.)

3.2.3 Frequency

The number of observations belonging to a certain event is called the frequency of that event in that sample. For example, Table 3.2 shows the frequency of number of accidents (in a month that varies from 0 to 11) during the 36-month time period.

3.2.4 Probability

Probability is a theoretical concept which can be defined in terms of frequency. Let the probability of an event be shown by P[A] where A is the event of interest. Then:

$$P[A] = f_n/n \text{ for } n \to \infty$$

where f is the frequency of the event of interest in the sample of size n. The event A can be the number of accidents in a given month as specified in Table 3.2. This can take values from 0 to 10 or more. Using the above definition, probabilities of events for A = 0 through A = 11 have been calculated in Table 3.2 and are shown in Figure 3.1.

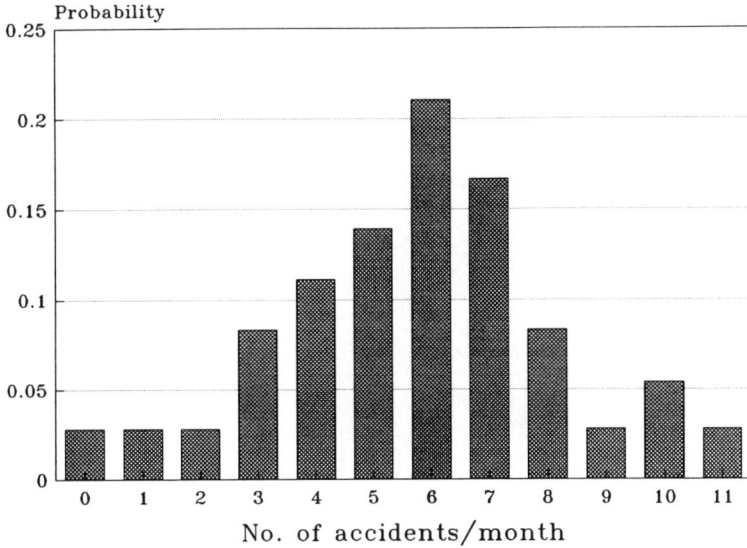

Figure 3.1. Probability chart for values calculated in Table 3.2.

3.2.5 Probability Density Function

Any event can be expressed in terms of a value or a set of values for the random variable. The probability of an event can then be expressed as the probability that a random variable will take on a value or set of values. Now, if this probability can be stated as a mathematical function of the random variable X, such that by knowing the value of X its probability could be calculated, this function is known as the probability density function for the random variable X.

As stated earlier, random variables can either be discrete or continuous and so are their corresponding probability density functions, i.e., discrete probability density function and continuous probability density function.

3.2.6 Discrete Probability Density Functions

For the random variable X, which is discrete, it may be possible to find a function of the type P(x) such that:

$$P[X = x_i] = P(x_i) \tag{3.1}$$

P [ ] is normally used to denote the probability of an event, and P(x) represents a function named P.

Example 3.1

Consider the following probability distribution function for the random variable X which can assume only 4 values of 1, 2, 3, and 4.

$$\text{For } x = 1,2,3,4 \quad P(x) = x/10 \quad\quad (3.2)$$
$$\text{Otherwise, } P(x) = 0$$

X may be considered to represent a number of accidents in a given period. Thus, we have

$$P[X = 1] = P(1) = 1/10$$
$$P[X = 2] = P(2) = 2/10$$
$$P[X = 3] = P(3) = 3/10$$
$$P[X = 4] = P(4) = 4/10$$

The sum of the above probabilities is given below.

$$P[X = 1] + P[X = 2] + P[X = 3] + P[X = 4] =$$
$$1/10 + 2/10 + 3/10 + 4/10 = 1$$

This function is shown in Figure 3.2.

It should be realized that the probability that X will take on any value other than those values comprising the sample space is zero. The above example sample space consists of only integer numbers from 1 through 4. It should also be noted that the sum of probabilities of all possible events is equal to 1.

F(x) is usually referred to as the cumulative probability distribution function and is given by the following expression:

$$F(x) = P[X \leq x] \quad\quad (3.2)$$

Thus, we have

$$F(1) = P[X \leq 1] = P(1) = 1/10$$

$$F(2) = P[X \leq 2] = P(1) + P(2) = 1/10 + 2/10 = 3/10$$

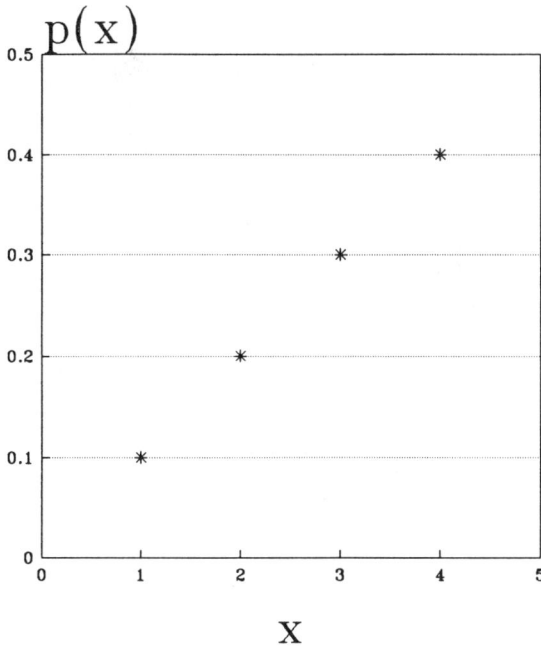

Figure 3.2. Graph of p(x) as a function of x.

$$F(3) = P[X \le 3] = P(1) + P(2) + P(3) =$$

$$1/10 + 2/10 + 3/10 = 6/10$$

$$F(4) = P[X \le 4] = 1/10 + 2/10 + 3/10 + 4/10 = 1$$

3.2.7 Continuous Probability Distribution Functions

For a continuous random variable, the probability of the random variable being exactly equal to a given value is zero. For continuous random variables, the probability is stated for an interval, i.e., $P[x_1 \le X \le x_2]$.

The cumulative distribution function in this case becomes:

$$F(x) = P[X \le x] \tag{3.3}$$

This function can be used to determine the probability of the event $x_1 \le X \le x_2$. A continuous probability density function of X, f(x) has the following properties:

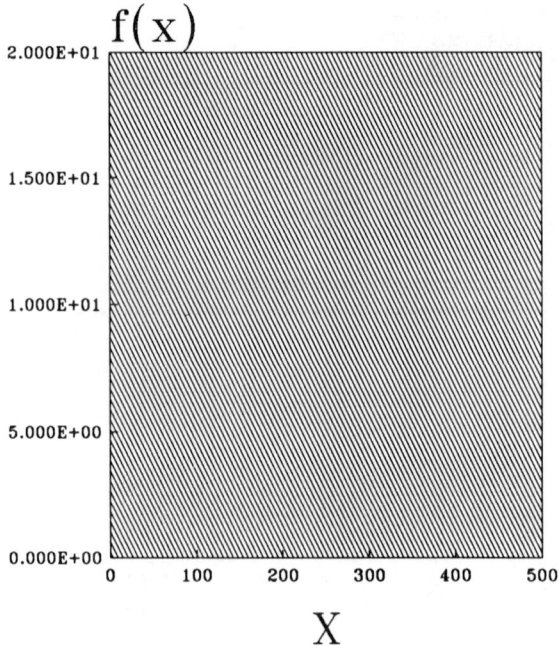

$$f(x)$$

Figure 3.3. Uniform probability density function.

1. f(x) is positive for all values of x.
2. The area under the curve of function f(x) between any two points is equal to the probability of the random variable falling between those two points.
3. The total area under the curve f(x) must equal to 1.

Example 3.2

In this example, consider the continuous random variable, lost-time due to a certain type of injury. Assume that the time lost due to this type of injury was anywhere between 0 to 500 h. Further, assume that no one interval of time has shown a higher probability of occurrence than any other.

Thus, the probability density function is

$$\text{For } 0 \leq x \leq 500 \qquad f(x) = 1/500 \qquad (3.4)$$
$$\text{Otherwise} \qquad f(x) = 0$$

This function is represented by Figure 3.3.

All the three properties of a continuous density function are satisfied by the diagram shown in Figure 3.3.

In our example, the probability of an accident causing lost hours between t_1 and t_2 is just 0.002 ($t_2 - t_1$). For example, the probability of an accident causing lost work hours between 200 and 250 is 0.1.

It is to be noted that many random variables exhibit a concentration of probability towards some midpoint. Distributions of this nature often fall into a class of distribution called "normal".

3.2.8 Characteristics of Random Variables

We do not know the value that a random variable may take, but there are quantitative measures by which a random variable can be identified. Following are some of these measures:

3.2.8.1 *Expected Value*

The expected or mean value of a random variable may be considered to be a value around which all possible values of the random variable are clustered.

For discrete random variables x_1, x_2, . . . , x_n the expected value is given by

$$\mu = E(X) = \sum_{i=1}^{n} x_i p(x_i) \qquad (3.5)$$

where $E(X)$ is the expected value or mean, x_i is one point in sample space, and $p(x_i)$ is the probability of x_i.

Example 3.3
Find the expected value of the random variable for the following probability density function:

$$\text{For } X = 1,2,3,4 \qquad P(x) = x/10 \qquad (3.6)$$
$$\text{Otherwise} \qquad P(x) = 0$$

Thus, we have $x_1 = 1$, $x_2 = 2$, $x_3 = 3$, and $x_4 = 4$, and the corresponding values of the probability density function are

$$p(x_1 = 1) = 1/10, \; p(x_2 = 2) = 2/10, \; p(x_3 = 3) = 3/10, \text{ and}$$

$$p(x_4 = 4) = 4/10$$

The expected value is

$$\mu = E(X) = \sum_{i=1}^{4} p(x_i) x_i$$

$$= 1(1/10) + 2(2/10) + 3(3/10) + 4(4/10) = 3$$

Similarly, the expected value for a continuous random variable is given by

$$\mu = E(X) = \int_{-\infty}^{+\infty} xf(x)\,dx \tag{3.7}$$

where $f(x)$ is the density function of the random variable X.

Though all the values of a random variable may not be known yet, it is possible to calculate the estimated mean, $\bar{x}$, of the distribution by taking a random sample of n measurements of the variable from its population and computing:

$$\bar{x} = \sum_{i=1}^{n} x_i/n \tag{3.8}$$

3.2.8.2   Variance

Let us consider 2 random variables A and B, and let the sample space for A consist of 6 values (5, 6, 7, 8, 10, 12) and that for B (7.5, 7.9, 8.0, 8.1, 8.2, 8.3). The mean values of A and B are as follows:

$$\mu_A = E(A) = 8$$
$$\mu_B = E(B) = 8$$

Note that although both random variables have the same mean, they are not similar. By looking at the sample space, it can be seen that points in sample space A are more dispersed than those in B. Dispersion is often measured by the deviation of points in the sample space from the mean and is known as variance.

The variance of a discrete random variable X having density function $f(x)$ is defined by

$$\sigma^2 = \sum_{i=1}^{n} (x_i - E(x))^2 f(x_i) \tag{3.9}$$

Summation sign is replaced by an integral for a continuous distribution.

Since all possible values of X are not known, it is necessary to estimate the true parameter σ^2 with an estimator S^2. Given a value of n measurements of the random variable, belonging to any distribution, the value of S^2 can be calculated from the following relationship:

$$S^2 = \frac{1}{n-1} \sum_{i=1}^{n} (x_i - \bar{x})^2 \qquad (3.10)$$

Using the above relationship, we get the following expression for standard deviation:

$$S = \sqrt{\frac{1}{n-1} \sum_{i=1}^{n} (x_i - \bar{x})^2} \qquad (3.11)$$

3.3   SELECTIVE PROBABILITY DISTRIBUTIONS

3.3.1   Poisson Distribution

The probability density function for this distribution is

$$P[X = k] = \frac{e^{-\lambda}\lambda^k}{k!}, \quad k = 0,1,2,\ldots \qquad (3.12)$$

where $k! = 1 \cdot 2 \cdot 3 \cdot \ldots \cdot k$.

This density function approximates situations in which an event takes place λ times, on the average, during a particular period. Number of accidents occurring in a given period of time are generally treated as a Poisson process.

This distribution has only one parameter λ which is equal to the mean of the random variable.

The mean and variance of the distribution are as follows:

$$E(X) = \lambda \text{ and } \sigma^2 = \lambda$$

Example 3.4

Number of accidents occurring during the previous one month was 1. What is the probability of having 0, 1, 2, 3, 4, 5, and 6 accidents in the coming month?

Utilizing the given data in Equation 3.12 yields:

$$P[x = k] = \frac{1^k}{k!} e^{-1} = \frac{e^{-1}}{k!}$$

$$P[x = 0] = \frac{1^0}{0!} e^{-1} = .368$$

$$P[x = 1] = \frac{1^1}{1!} e^{-1} = .368$$

$$P[x = 2] = \frac{1^2}{2!} e^{-1} = .184$$

$$P[x = 3] = \frac{1^3}{3!} e^{-1} = .061$$

$$P[x = 4] = \frac{1^4}{4!} e^{-1} = .015$$

$$P[x = 5] = \frac{1^5}{5!} e^{-1} = .003$$

$$P[x = 6] = \frac{1^6}{6!} e^{-1} = .001$$

$$P[x \leq 1] = P[x = 0] + P[x = 1] = .368 + .368 = .736$$

A graph of this distribution is shown in Figure 3.4. $P[X \leq 1]$ denotes the probability of having the utmost 1 accident.

Tables showing the summation of the terms of Poisson distribution for large values of λ and k are generally available in standard texts on probability and statistics. Table A.1 in Appendix A gives selected values of the Poisson cumulative distribution function.

3.3.2 Normal Distribution

The normal distribution is one of the commonly used distributions, and its properties are well known. Under the assumption of normality for a given set of data holds, various statistical tests may be conducted using this distribution. In turn, to determine normality of a given set of data, the tests such as the Kolmogrov-Smirnov[1] may be used.

Probability

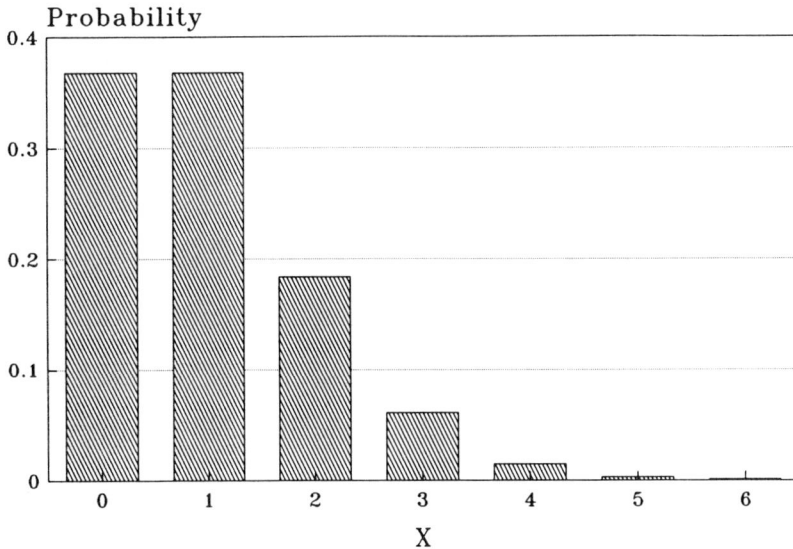

Figure 3.4. Graph of Poisson distribution.

The probability density function of the normal distribution is defined by

$$f(x) = \frac{1}{\sqrt{2\pi}\sigma} \exp[-(x - \mu)^2/(2\sigma^2)],$$

$$\text{for } -\infty < X < \infty, \text{ and } \sigma > 0 \tag{3.13}$$

where μ is the mean of the distribution and σ is the standard deviation.

If a random variable has a normal distribution and its mean and standard deviation are known, any measurement from that distribution will be within ± 1 standard deviation (SD) of the mean approximately 68.4% of the times, ± 2 SD 95% of the times, and ± 3 SD 99% of the times as shown in Figure 3.5. If we substitute $z = (x - \mu)/\sigma$, in Equation 3.13, the probability density function for z becomes

$$f(z) = \frac{1}{\sqrt{2\pi}} \exp(-z^2/2), \text{ for } -\infty < z < +\infty \tag{3.14}$$

This function is known as standardized normal density (having mean 0 and variance 1) function. Table A.2 in Appendix A, showing the values of the standardized normal function, can be used for calculating

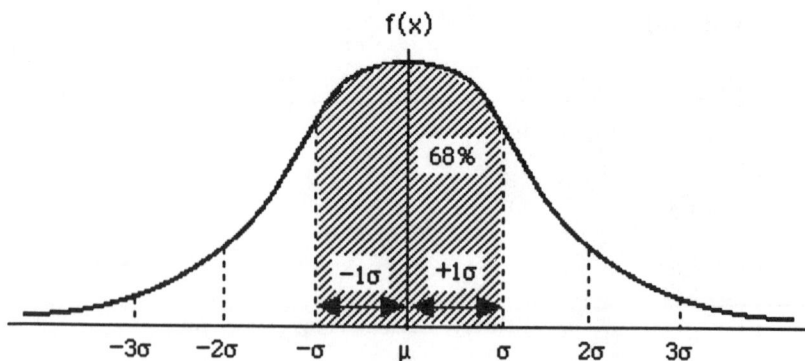

Figure 3.5. Normal distribution probability structure.

the probability of a normally distributed variable departing from the mean by more than z standard deviations.

Example 3.5

Suppose that the number of accidents in a company over a 1 year period are normally distributed with a mean of 12 and a SD of 2, calculate the probability that 18 accidents in a given year are due to chance alone. Thus, we get

$$z = (18 - 12)/2 = 3$$

Using Table A.2 we get the probability value .9987. Hence, there is only a chance of .0013 (1 − .9987) or a 0.13% chance that there would be 18 more accidents in the year.

3.3.3   Distribution of Sample Means

The mean of the sample means is the mean of the population of individual values from which the samples are drawn.

The standard deviation of the distribution of sample means equals $1/\sqrt{n}$ times the standard deviation of the population individual values. The symbol n denotes sample size.

The form of the distribution of sample means approaches the form of a normal probability distribution as the size of the sample is increased.

3.3.4 The Normal Distributions of X

In the example given earlier, it was necessary that the random variable, i.e., the number of accidents per year, was normally distributed. There may be many occasions when the random variable may not be normally distributed. For such cases we may, instead of testing an individual value from the underlying population, test the value of a sample mean of size n. In such a case, $\overline{X}$ (i.e., $\overline{X} = \Sigma_{i=1}^{n} x_{i/n}$) becomes a random variable. If n is greater than 30 then $\overline{X}$ may be considered to be normally distributed. Figure 3.6 shows three distributions of populations and distributions of sample means for n = 2,5 and 30. The distributions of $\overline{X}$ when n is 30 look like normal distributions.

Example 3.6

Accidents in a textile industry in a given province (state) have a mean rate of 5 accidents per week, with a SD of 0.60. What is the probability that the mean accident rate of a random sample of 50 textile factories in the province will be between 5.1 and 5.2? In this example, we have

$$\mu_x = 5 \text{ and } \sigma_x = 0.6$$

From the sampling distribution of $\overline{X}$, $\mu_{\overline{x}} = \mu_x = 5$, the standard deviation of the mean

$$\sigma_{\overline{x}} = \sigma_x / \sqrt{n} \tag{3.15}$$

For n = 50, the mean of $\overline{X}$ = E $(\overline{X})$ = 5.0. Thus, the standard deviation of the sample mean $(S_{\overline{x}})$ = $0.6/\sqrt{50}$ = 0.0849

$$Z = \overline{X} - E(\overline{X})/S_{\overline{x}} \tag{3.16}$$

For $\overline{x}$ = 5.10, we have z = (5.10 − 5.0)/0.0849 = 1.18, and for $\overline{x}$ = 5.2, z = (5.2 − 5.0)/0.0849 = 2.36. Probability that the mean accident rate is between 5.1 and 5.2 is P (z ≤ 2.36) − P (z ≤ 1.18). Using normal tables we get the following value for the probability:

$$.4909 - .3810 = .1099 \doteq .11$$

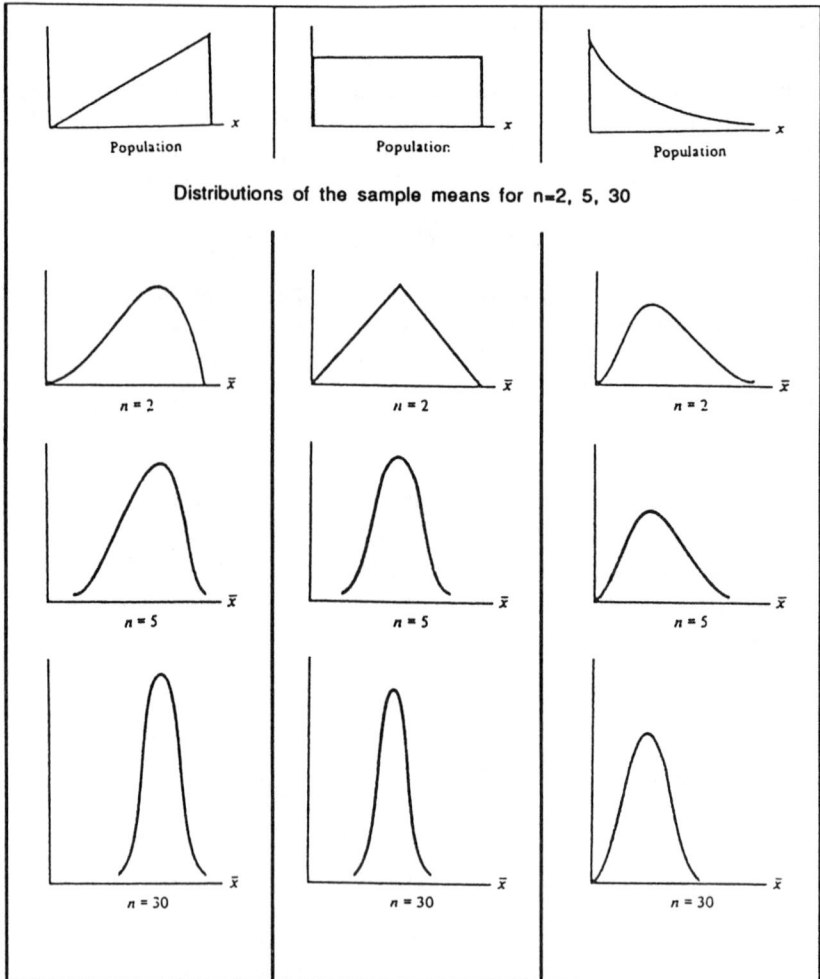

Distributions of the sample means for n=2, 5, 30

Figure 3.6. Distributions of the populations.

So there is an 11% chance that the mean accident rate of the 50 textile factories selected at random within the province will be between 5.1 and 5.2.

3.3.5   The t Distribution

In the earlier section, it was assumed that the population standard deviation is known, and the sample size is equal or greater than 30. In practice, this may not be so. To handle such situations, application of t distribution is useful:

$$t = (\bar{x} - \mu)/(s/\sqrt{n}) \tag{3.17}$$

where t = value of random variable said to have a t distribution
$\bar{x}$ = mean of the sample of size n
μ = population mean
s = sample standard deviation

t statistic has a different distribution for each value of n. Values of t for n = 2 to 30 are given in Table A.3 in Appendix A.

Example 3.7

The safety officer of a large industrial plant sought to determine the mean daily amount of sulfur oxide emitted by the plant. As the measurement costs were high, only a random sample of 9 d was obtained. These were 95, 98, 92, 84, 105, 92, 110, 86, and 98 K.G.s/d.

The average emissions in the past have been 90 K.G./d. What conclusions can be drawn concerning the shift in the amount of emissions?

$\bar{X} = (95 + 98 + 92 + 84 + 105 + 92 + 110 + 86 + 98)/9 = 95.56$

Similarly, we have

$$S = \sqrt{\sum_{i=1}^{n} (x_i - \bar{x})^2/(n - 1)} = 8.37 \tag{3.18}$$

Substituting the above data in Equation 3.15 yields

$$S_{\bar{x}} = 8.37/\sqrt{9} = 2.79$$

Using the above results and the given data in Equation 3.17 lead to

$$t = (95.56 - 90)/2.79 = 1.99$$

The above result indicates that the sample value deviates from μ = 90 by 1.99 σ units.

Using "t" tables, we find that the probability of the random variable being 1.99 σ or less σ units from μ = 90 when n = 9 is between 0.95 and 0.975. By interpolation, this is calculated to be 0.96.

This means that there is only a 4% chance that the mean of the sample has increased, which for the time being may not necessitate any action except taking further samples at a later time.

3.4   TESTS OF HYPOTHESES

Example 3.7 has shown the use of statistical estimation to draw conclusions. In this section, we shall start with an assumed value of a population parameter, then we shall use sample evidence to decide whether the assumed value is unreasonable and should be rejected, or whether it should be accepted. The assumptions that we make about the population parameters are known as hypotheses. In each test, there will be two hypotheses. One is known as the null hypothesis (H_0), and the other is called the alternate hypothesis (H_a). These are stated in a form such as:

H_0:   5 accidents per unit time
H_a: 5 accidents per unit time

Hence the null hypothesis is that the population mean is greater than or equal to 5, and the alternate hypothesis is that the population mean is less than 5.

3.4.1   Type I and Type II Errors and Steps for Performing a Test of Hypothesis

Sample evidence is used to test H_0; if that sample evidence indicates that H_0 has only a small probability of being true (i.e., a large probability of being false), one can say that H_0 is rejected. In case the sample evidence is not convincing that H_0 is unreasonable, then one accepts H_0.

It is obvious from the above that one cannot be sure of the conclusions, as these are based upon sample evidence. A conclusion is correct if a true H_0 is accepted or if H_a is rejected. We shall make an error if a true H_0 is rejected or if H_a is accepted. Rejecting a true H_0 is called a Type I error, and accepting a false H_0 is known as a Type II error.

The general procedure that is usually followed for testing a hypothesis is composed of the following steps:

1.   State the problem and formulate an appropriate null hypothesis H_0 with an alternative hypothesis H_a.

2. Decide a level of significance (i.e., probability of rejecting a true H_0).
3. State the test-statistic to be used, e.g., the normal distribution, the t-distribution, etc.
4. Compute the value of the test-statistic from the sample values in order to decide whether to accept or reject the null hypothesis H_0.
5. Determine the region of rejection using tables given in Appendix A. Formulate the decision rule as given below:

   - If the value of the test-statistics falls in the rejection region, reject the hypothesis.
   - Accept the hypothesis otherwise.

3.4.2 Forms of Hypothesis

Let us assume μ_1 as the observed population mean. Three possible null hypotheses regarding a population mean are stated below.

1. $H_0: \mu \leq \mu_1$
   $H_a: \mu > \mu_1$

2. $H_0: \mu \geq \mu_1$
   $H_a: \mu < \mu_1$

3. $H_0: \mu = \mu_1$
   $H_a: \mu \neq \mu_1$

Tests (1) and (2) lead to one-tail test (i.e., (1) is a right-tail test, and (2) is a left-tail test). On the other hand, Test (3) is a two-tail test. In each case, the test is to be conducted by obtaining a random sample of size n and computing its mean, $\overline{X}$. And then $\overline{X}$ is used in computing a test statistic. Depending on the value of the test statistic, H_0 is accepted or rejected.

3.4.3 The Test Statistics

Two types of test statistics are described below.

1. Use the Z test-statistics when (1) the population is normal and σ_x is known, or when the sample size is greater than 30. Thus, we have

Table 3.3 Rejection Regions for H_0

When the alternative hypothesis is	The rejection region will be
H_1: $\mu < \mu_0$, (one-sided)	$Z < -Z_{\alpha}$[1]
H_1: $\mu > \mu_0$, (one-sided)	$Z > Z_{\alpha} - Z_{\alpha/2}$
H_1: $\mu \neq \mu_0$, (two-sided)	$Z < -Z_{\alpha/2}$ or $Z > Z_{\alpha/2}$

[1] Z_{α} represents the value of Z at an α significance level (see normal distribution table).

Table 3.4 Rejection Regions for H_0

When the alternative hypothesis is	The rejection region will be
H_1: $\mu < \mu_0$,	$t < -t_{\alpha(n-1)}$,[1]
H_1: $\mu > \mu_0$,	$t > -t_{\alpha(n-1)}$,
H_1: $\mu_1 \neq \mu_2$	$t < t_{\alpha/2(n-1)}$ or
	$t > t_{\alpha/2(n-1)}$

[1] $t_{\alpha}(\ )$ is obtained from the t-table for desired significance level and the degree of freedom.

$$Z = (\overline{X} - \mu_h)/\sigma_{\overline{x}} \qquad (3.19)$$

where $\sigma_{\overline{x}}$ = $\overline{X}$ standard deviation

$\quad\quad \sigma_{\overline{x}}$ = $\sigma/\sqrt{n}$

$\quad\quad \sigma$ = population standard deviation

$\quad\quad \mu_h$ = hypothesized population mean

The rejection regions for H_0 corresponding to different alternative hypotheses are given in Table 3.3.

2. Use the t statistics when the population is normal, with a specified value for its unknown mean, and σ^2 is not known or $n \leq 30$. Thus, we have

$$t = [(\overline{x} - \mu_h)\sqrt{n}]/s \qquad (3.20)$$

where

$$\overline{x} = (1/n) \sum_{i=1}^{n} x_i, \quad s^2 = (1/(n-1)) \sum_{i=1}^{n} (x_i - \overline{x})^2 \text{ and}$$

$\quad\quad \mu_h$ = hypothesized mean

The critical regions for H_0 corresponding to different alternative hypotheses are given in Table 3.4.

Example 3.8

Assume that the data regarding noise level at a given workplace in a company is 85 db with a standard deviation of 4 db. A sample of 36 measurements yielded a mean of 83.6 db. Test the hypothesis that H_0: μ = 85 db against H_a: μ < 85 db at α = 0.01. Using the above given data, we solve this example as follows:

1. We formulate our null and alternative hypotheses as given below:

$$H_0: \mu = 85$$
$$H_1: \mu < 85$$

2. We set the level of significance at α = 0.01.
3. Since the sample size is greater than 30, we shall use Z test-statistic. Therefore, the test-statistic to be used is

$$Z = [(\bar{x} - \mu_h) \sqrt{n}]/\sigma_{\bar{x}} \qquad (3.21)$$

4. The calculated value of Z = -2.1.
5. The rejection region will be Z < $-Z_\alpha$. However, in our case we have Z = -2.1 and $-Z_\alpha$ = -2.326, therefore, we accept H_0.

To show the use of t-statistics, let us consider a similar problem when the noise level was 85 db on the average. Some modifications to lower the noise level of the process were carried out. After the modifications, the following sample of noise measurements was taken:

82.1, 85.3, 86.3, 84.3, 85.2, 82.2, 81.3, 84.0, 86.0

For α = .05, find out if the noise level has, in fact, been significantly reduced.

Since the number of measurements in the sample is less than 30, we will use the t-statistics as follows:

1. We state the null hypothesis and the alternative hypothesis as given below:

$$H_0: \mu = 85$$
$$H_a: \mu < 85$$

Table 3.5 Data for Two Similar Departments

	Average no. of lost-time accidents ($\bar{x}$)	Sample standard deviation (s)
Department 1	16.1	4.3
Department 2	14.3	5.1

2. In this, we have $\alpha = 0.05$.
3. Since the number of measurements is less than 30, we will use the t-statistics:

$$t = [(\bar{x} - \mu_h)\sqrt{n}]/s \qquad (3.22)$$

In our case, $\bar{x} = 84.09$ and $s = 1.82$.
4. The calculated value of t-statistics is $= -1.5$.
5. The critical region would be $t < -t_{0.05(8)} = -1.860$. Thus, we accept the null hypothesis because the calculated value of t is greater than the tabulated value of $t_{0.09(8)}$.

It is worth mentioning that if the alternative hypothesis was $\mu > 85$, then the critical region would be $t > t_{0.05(8)}$. In this case, we would have rejected the null hypothesis.

3.4.4 Hypothesis Concerning Two Means

Let us assume for a moment that we are interested in comparing the accident records of two similar departments in a given plant. Number of lost-time accidents are used as the measurement. Under such condition, we have the data given in Table 3.5 covering a period of 36 months. The following steps describe this hypothesis:

1. Let μ_1 and μ_2 be the mean of the lost-time accidents of Department 1 and Department 2, respectively. Here, our null and alternative hypotheses are defined below.

$$H_0: \mu_1 = \mu_2$$
$$H_a: \mu_1 > \mu_2$$

2. Assume $\alpha = 0.05$.

3. Since the sample sizes are sufficiently large and the population variances are unknown, we take the given standard deviations as estimates of σ_1 and σ_2.

$$z = (\bar{x}_1 - \bar{x}_2)/\sqrt{s_1^2(n_1 - 1) + s_2^2/(n_2 - 1)} \qquad (3.23)$$

4. The value of Z is calculated and equals 1.5963.
5. The rejection region for $\alpha = 0.05$ is $Z > Z_{0.05} = 1.645$ our calculated Z is less than the tabulated $Z_{0.05}$. Therefore, we conclude that at 0.05 level of significance, Department 2 does not have significantly different lost-time accidents than Department 1.

For cases where $n < 30$, t-statistics[1] can be used:

$$t = \frac{(\bar{x}_1 - \bar{x}_2)\sqrt{n_1 n_2(n_1 + n_2 - 2)}}{\sqrt{n_1 - 1)\, s_1^2 + (n_2 - 1)\, s_2^2\, (n_1 + n_2)}} \qquad (3.24)$$

3.5 CONTROL CHARTS

A control chart is a diagram which helps in making judgements regarding the state of a process being in control or not. In our case, the process may be the frequency of accidents, severity of accidents, or some other indirect indicators of safety.

A record of past accidents may be used to construct control charts for monitoring the present state of safety and also for forecasting the same. The formulas used are based on Poisson distribution, as most accident data assume its shape. The period used can be the previous 12 months or any other period that would be representative.

The following formulas are used in developing control charts:

$$\bar{p} = \text{the sample mean} \qquad (3.25)$$

$$\text{Accident Frequency} = \frac{\text{Total number of disabling injuries}}{\text{total exposure time}} \times \text{base factor} \qquad (3.26)$$

$$\text{Standard Deviation } (\sigma_p) = [(p \times \text{base factor})/(\text{exposure interval})]^{1/2} \qquad (3.27)$$

Table 3.6 Disabling Injury Data for a Twelve Month Period for an Organization

Month	No. of disabling injuries	Exposure in employee hours	Disabling injuries per 200,000 employee hours
Jan.	80	150,000	106.66
Feb.	80	150,000	106.66
Mar.	75	150,000	100.00
Apr.	50	150,000	66.66
May	60	150,000	80.00
June	70	150,000	93.33
July	55	150,000	73.32
Aug.	65	150,000	86.67
Sep.	80	150,000	106.66
Oct.	75	150,000	100.00
Nov.	70	150,000	93.33
Dec.	65	150,000	86.67
Total	825	1,800,000	1,099.96

Base factor is usually considered to be equal to 200,000. This is based on the assumption that there are 100 full-time employees, each working for 50 weeks per year and 40 hours per week.

Example 3.9

Let us assume that Table 3.6 shows the number of disabling injuries recorded each month in a company employing 75 full-time workers (each worked for 50 weeks per year, and 40 hours per week). Using these data, construct a control chart.

Using the data specified in Table 3.6, we get the following results:

average injuries per 200,000 employee-hour exposure $(\bar{p})$
$$= *(825/1800,000) \times 200,000 = 91.67$$
$$\sigma_p = [\ (\bar{p} \times \text{base factor})/(\text{exposure interval})\ ]^{1/2} = 11.06$$

For constructing control limits with 95% confidence that all accidents will fall within $\pm 2\ \sigma_p$, we get

Upper Control Limit (UCL) $= \bar{p} + 2\sigma_p = 91.67 + 2(11.06) = 113.79$

and

Lower Control Limit (LCL) $= \bar{p} - 2\sigma_p = 91.67 - 2(11.06) = 69.55$

The accidents of each month can now be charted to determine if they lie within expected limits. Figure 3.7 shows the control chart for the data.

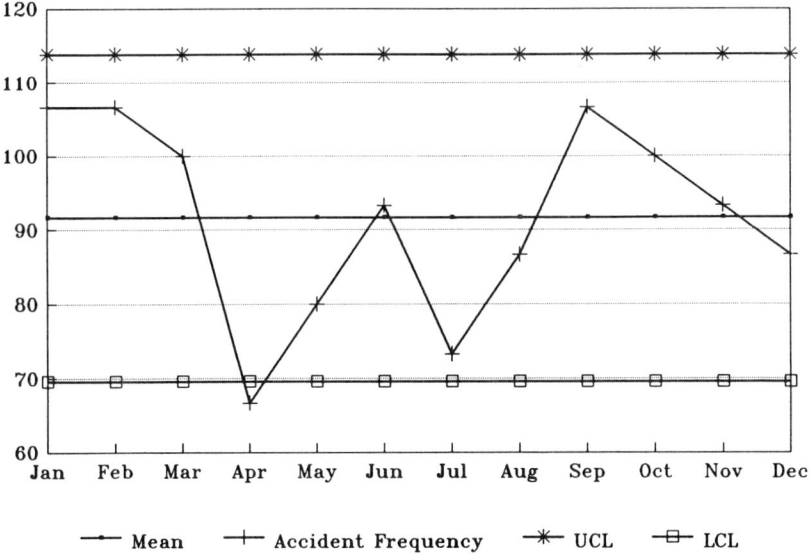

Figure 3.7. Control chart for Table 3.6 data.

By examining this chart, we may say with 95% confidence that 90% of the monthly accident figures fall below the UCL. Such a control chart can also be used to determine goals and to measure results after implementation of a safety improvement program. As a goal, we might like to add that in the coming year, we shall attempt to reduce accidents to or below the mean, say, ($\bar{p}$ for the current year) of this year. If the new program was implemented, then we can effectively monitor the progress.

In this example, we assumed that exposure in employee hours was the same for all the months. This means that none of the employees was absent during the entire year. In fact, it may not be the case as a total number of hours worked may vary from month to month.

Example 3.10

In this example, the number of hours vary from month to month. The UCL for each month for 95% confidence have been computed as given in Table 3.7. From these values, it is easily seen that the process is well in control. Using the data specified in Table 3.7, we get the following result:

average injuries per 200,000 employee-hour exposure ($\bar{p}$) =
   $(29/368,600) \times 200,000 = 15.74$

Table 3.7 Upper Control Limit (UCL) Values for the Situation where the Number of Hours Vary from Month to Month

Month	No. of disabling injuries (c)	Exposure in employee hours	Disabling injuries per 200,000 employee hours	2σ	UCL
Jan.	2	32,300	12.4	19.74	35.48
Feb.	2	26,500	15.1	21.80	37.54
Mar.	2	25,200	15.9	22.35	38.09
Apr.	0	33,000	0	19.53	35.27
May	1	34,300	5.8	19.16	34.90
June	1	34,800	5.8	19.02	34.76
July	3	24,600	24.4	22.62	38.36
Aug.	4	25,000	32.0	22.44	38.18
Sep.	3	29,100	20.6	20.80	36.54
Oct.	3	37,200	16.1	18.40	34.14
Nov.	4	37,000	21.6	18.45	34.19
Dec.	4	29,600	27.0	20.63	36.37
Total	29	368,600	196.7		

Using $\bar{p}$ as the center line of the control chart, UCL for each month was calculated.

Table 3.7 indicates that none of the values are statistically out of control. It may be noticed that at $\bar{p}$ = 15.74, the distribution is not uniform. Further analysis by calculating the probability of that frequency occurring in the sample size (exposure in employee hours) found in that month may be carried out as follows:

For the month of August

$$\text{number of disabling injuries} = 4$$

$$\text{number of hours exposure} = 25,000$$

$$\bar{p} = 15.74$$

$\bar{p}$ is the average frequency per 200,000 working hours; however, the plant did not work for that duration, so the expected average number of accidents would be somewhat lower. Thus, we get expected number of accidents ($\bar{p}$ for sample)

$$= \text{universe average} \times \text{sample size}$$

$$= 15.74 \times 25000/200,000$$

$$= 1.97$$

The probability of having 4 or more accidents when the mean expected number is 1.97 is calculated using the Poisson probability distribution function as follows:

$$P(c \geq 4) = P(4) + P(5) + P(6) + P(7) + \ldots$$

This value from the standard Poisson distribution tables is

$$P(c \geq 4) = 0.053$$

This means, in the long run, that we expect 4 or more accidents 5.3% of the time.

It is advantageous to calculate the probabilities of the number of accidents being equal or beyond the actual number of disabling injuries. With the knowledge of this, one can start looking for assignable causes of disabling injuries for those cases where the probability is less than, say, .05.

3.6 PROBLEMS

1. Describe the following terms:
   A. Probability
   B. Discrete random variable
   C. Continuous random variable

2. Discuss the difference between the probability density function and the probability cumulative distribution function.

3. Assume that at a major industrial plant, times to accident are exponentially distributed. Obtain an expression for the mean time to accident.

4. For the above, assume that the mean time to accident is 1500 working hours. Calculate the value of the accident rate.

5. Assume that the number of accidents occurring at a plant during the previous months was two (per month). Calculate the probability of having 0, 1, and 2 accidents in the coming month.

6. Assume that in one of the states, the electronic industry has an average rate of 4 accidents per week with a standard deviation of 0.65. Compute the probability that the average accident rate of a random sample of 40 electronic factories in the entire state will be between 4.2 and 4.5.

7. Assume that the noise level in a certain area of a factory is 100 db with a standard deviation of 5 db. A sample of 40 measurements yielded an average of 95 db. Test the hypothesis that H_0: $\mu = 100$ db against H_a: $\mu < 100$ db at $\alpha = 0.05$.

REFERENCES

1. Miller, I. and Freund, J.E., *Probability and Statistics for Engineers,* Prentice-Hall, Englewood Cliffs, NJ, 1977.
2. Dey, K.A., *Practical Statistical Analysis for the Reliability Engineer,* Rome Air Development Center, Griffiss Air Force Base, New York, 1983.
3. Brown, D.B., *Systems Analysis and Design for Safety,* Prentice-Hall, Englewood Cliffs, NJ, 1976.
4. Dhillon, B.S., *Quality Control, Reliability, and Engineering Design,* Marcel Dekker, New York, 1985.
5. Dhillon, B.S., *Reliability Engineering in Systems Design and Operation,* Van Nostrand Reinhold, New York, 1983.
6. Dhillon, B.S., *Robot Reliability and Safety,* Springer-Verlag, New York, 1991.

CHAPTER 4

Safety Behavior Sampling

4.1 INTRODUCTION

Accidents have been called members of a broad family of errors of people.[1] Two causes of errors are error-committing characteristics and error-provocation situations. One way to reduce these errors is to urge people to minimize their error-committing characteristics. This may be achieved by providing necessary feedback to them concerning errors. Figure 4.1 presents some of the major factors influencing errors in human performance. Errors made by the workers are often known as unsafe acts, and the existence of error provocation situations is called unsafe conditions. Heinrich[2] has stated that roughly 88% of all accidents are caused by the unsafe acts of people, 10% by unsafe conditions, and 2% by the act of God. This breakdown may not be universally accepted, as is evidenced from Figure 4.2. The data used in this figure were obtained from a large company employing between 2000 to 2200 workers. However, Heinrich's classification definitely provides basis for the broad categorization of accident causes.

This chapter is concerned with safety behavior sampling — a technique for measuring unsafe acts.

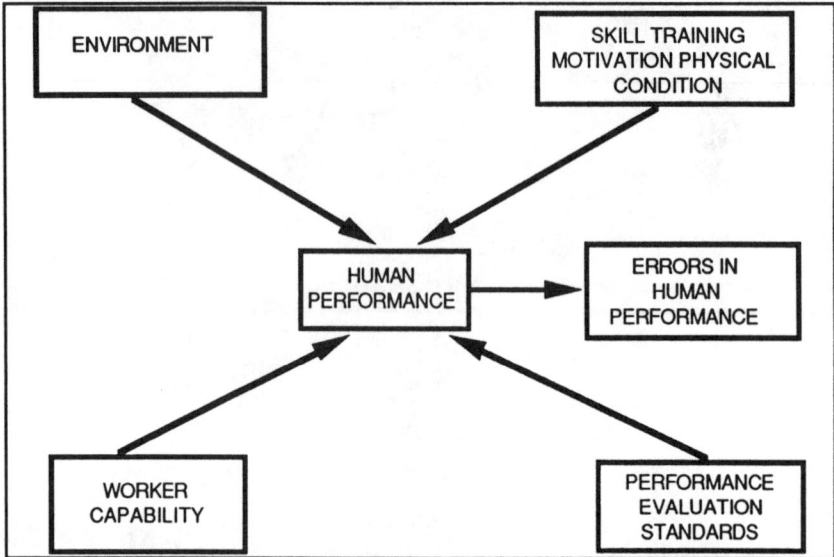

Figure 4.1. Factors influencing errors in human performance.

4.2 FUNDAMENTALS OF SAFETY BEHAVIOR SAMPLING

Safety behavior sampling is based on the same principles as work sampling used in industrial engineering for purposes such as establishing time standards.

The whole concept of work sampling is based on the laws of probability. For example, if we are dealing with a "process" which can be only in two states (i.e., on or off, yes or no, working or idle, safe or unsafe), we know the percentage of time the two states can occur must equal 100%. In a multiactivity study, each observation is in a binary state for each activity considered. In terms of probabilities, we can express this relationship as

$$p + q = 1$$

where p = the probability of a single observation in one state, say S for safe act

    q = (1 − p) = probability of no observation in state S

The above relationship can be extended to include n observations:

$$(p + q)^n = 1$$

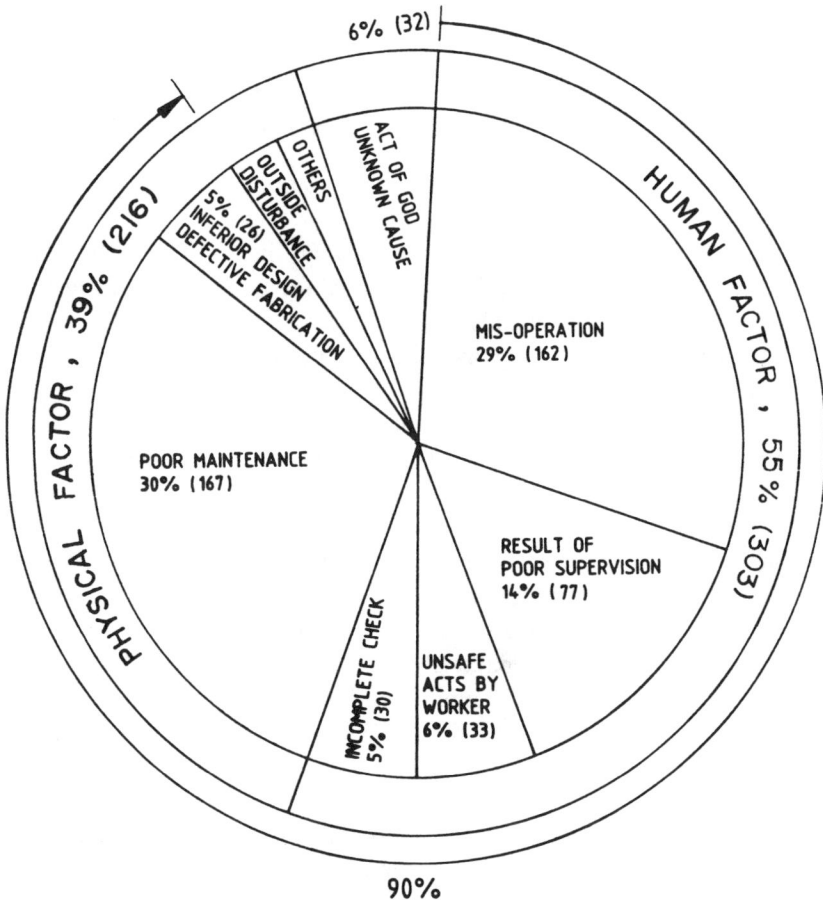

Figure 4.2. Accident cause classification (total number of accidents = 551). Figure in brackets denotes number of accidents.

where n is the number of observations in the sample. The expression can be expanded to obtain the probability that a certain number of observations will be in state S out of a total of n observations.

The distribution of probabilities resulting from the binomial expansion follows the binomial distribution. The mean of this distribution is equal to np and the standard deviation is given by $\sqrt{npq}$. As n becomes large, the binomial distribution takes on the properties of the normal distribution. This means that the binomial distribution can be closely approximated by the normal distribution for a large value of n when neither p nor q is close to zero. In order to use the normal approximation of the binomial distribution, we have to divide both the mean and the standard deviation by the sample size as follows:

$$\text{sample mean } = np/n = p \qquad (4.1)$$

$$\text{sample standard deviation } = \sqrt{npq}/n$$

$$= \sqrt{pq/n} \qquad (4.2)$$

$$= \sqrt{p(1 - p)/n}$$

Jewel,[3] Rockwell,[4] Schreiber[5], Meyer,[6] and Pollina[7] have demonstrated the usefulness of this sampling technique for evaluating unsafe behavior. In applying safety behavior sampling, it is assumed that the percent of time a worker working safely/behaving unsafely, can be determined.

In order to obtain a complete and accurate picture of safe/unsafe acts performed by the worker, it is necessary to continuously observe the worker and record data related to unsafe acts. Note that a sufficiently large sample must be obtained for representative results.

For a large number of observations, the resulting distribution approaches the shape of a normal curve.

Properties of a normal distribution are presented in Chapter 3. As stated in this chapter, approximately 95% of all observations fall within ±2 standard deviation (SD) limits and 99% within ±3 SD limits. It is emphasized that the normal distribution may only be assumed for a large number of observations. One must determine the adequacy of the number of observations in terms of the desired level of confidence. Generally, a confidence level of 95%, or within ±2 SD, is considered adequate for most safety behavior sampling studies.

This confidence level means that the conclusions will be representative of the true population 95% of the time.

In addition to the confidence level, the safety behavior sampling has another attribute called accuracy. Accuracy may be interpreted as the tolerance limit of the observations that fall within a desired confidence level. A 5% accuracy with 95% confidence level is the combination often used in safety behavior sampling. This means that 95% of the time within ±5% accuracy limit, the conclusions drawn on the basis of safety behavior sampling will be representative of the actual population.

Safety behavior sampling requires randomness of occurrences. This is achieved when each period of the workday is equally likely to be selected as the observation period.

4.3   PROCEDURE FOR SAFETY BEHAVIOR SAMPLING

4.3.1   Define Work Stations

This includes departments/units in an organization where safety behavior sampling is to be conducted: tool room, punch press section, stock and shipping, etc.

4.3.2   Prepare a List of Unsafe Acts

This list can be developed from plant accident records initially and modified later as appropriate. Plant accidents include all accidents, such as disabling injuries, recordable injuries, and first aid cases. However, the lists of unsafe acts contained in the National Safety Council's Accident Prevention Manual for Industrial Operations may also be used. A list of unsafe acts obtained from a critical Incident Study by Tarrants[8] (classified according to Modified ANSI Classification System) is given in Table B.1 in Appendix B. This list may provide useful reference for preparing a list of unsafe acts for the plant/ department to be studied. A specimen of Safety Behavior Sampling Worksheet listing unsafe acts is shown in Figure 4.3.

4.3.3   Conduct a Pilot Study

Prior to conducting a pilot study, one must carefully select times to observe worker behavior. These times must be selected randomly. The number of trial observation periods required depends upon the number of persons observed. For a guideline purpose, it is said that a sufficient number of trial observation periods should be selected so that the total sample size is at least 100. When planning trial observation periods, there are many methods available for arranging observation periods in a random pattern.[9] One such method is described below.

In this method, the first hour of the working day is identified by the numeral 1 and the second by the numeral 2 and so on. A table of random numbers is used to obtain a series of three digit figures, the first digit representing the hour of the working day and the next two the minutes. Numbers representing hours not in the working day, or impossible minute values, are discarded. A sufficient quantity is obtained to give the required observation times for each day of the study. There should always be a separate list for each day.

SAFETY BEHAVIOR SAMPLING

WORK SHEET

Department:_____   Type of activity:_____

Work Center:_____

UNSAFE ACTS					
1. Handling Hot Parts with unprotected hands					
2. Failure to wear proper safety glass					
3. Failure to wear proper safety glasses					
4. Improper lifting					
5. Carrying heavy load					
_____					
_____					
_____					
Total Unsafe Acts.					

Date:

Time:

Safety Analyst:

Figure 4.3. Safety behavior sampling worksheet.

Example 4.1

Assume that a plant operates from 8:00 a.m. to 5:00 p.m. In this case, the time 8:00 a.m. is designated by the first digit of the random number, 9 a.m. by 2, and so forth, down to 4 p.m. by 9. A short random number and its interpretation is given in Table 4.1.

For actual use, the list of observation times is to be arranged in time sequence.

Table 4.1 A Random Number Table and Its Interpretation

Random no.	Interpretation
907	4.07 p.m.
882	Impossible minute value — discard
544	12.44 p.m.
720	2.20 p.m.
838	3.38 p.m.
010	Impossible hour — discard
413	11.13 a.m.

Observer Training

The observer must be instructed to categorize worker behavior as being either safe or unsafe as defined by the entries of behaviors included in the unsafe acts list. He/she should make a trial run and practice how to decide instantaneously whether the observed behavior is safe or unsafe. In addition, the observer should be trained to determine whether the behavior of each worker is safe or unsafe at the time of each observation from a single observation point if observing workers from a given group are desired. When the observer is required to study a department or a plant, he/she should walk through and determine the proportion of workers behaving unsafely.

Calculation of Required Number of Observations

The number of observations required is based on data collected during the pilot study, the degree of accuracy required, and the given level of confidence.

During the pilot study, two items are recorded:

1. Total number of observations made (N_1)
2. Number of observations in which the unsafe behavior was observed (N_2)

Thus, the proportion of unsafe behavior is

$$N_2/N_1 = p \qquad (4.3)$$

If S = desired accuracy
   N = total number of observations required, and
   K = the value obtained from standardized normal tables for a given level of confidence, then

$$N = (K/S)^2 p(1 - p) \qquad (4.4)$$

For a given level of confidence K, the value of K is read from the standardized normal tables. For 95% confidence, k is approximated as 2, and for 99% confidence, k is taken as 3.

Thus, for a 95% confidence, Equation 4.4 becomes

$$N = (4/S^2)p(1 - p) \qquad (4.5)$$

Example 4.2

Assume that a pilot study generated the following results:
Total observations $= 120 = N_1$
Unsafe observations $= 35 = N_2$
Thus, we have

$$p = 35/120 = .029$$

Now, assume that we want to calculate N for a 10% accuracy and 95% confidence level. Thus, substituting the above given values into Equation 4.4 yields:

$$N = (2/0.1)^2 \, 0.29(1 - 0.29) \simeq 82$$

This means a minimum of 82 observations are needed to produce satisfactory results.

4.3.4 Conducting Additional Observations

Carry out the actual study by making $(N - N_2)$ number of random observations and recording the behavior according to safe and unsafe classifications.

After conducting the additional observations, calculate S by using the following formula:

$$S = \sqrt{p(1 - p)/N} \qquad (4.6)$$

In the case of the above example, we have $K = 2$ and $N = 82$.

If $S \leq .1$, the results obtained provide us the desired accuracy and confidence level; in case $S > .1$, calculate new N and start making the observations as explained above.

This study should be repeated once a week for a series of 6 to 12 weeks. Number of repetitions depends upon the sample size N, the number of persons observed, and the available manpower to undertake the study.

4.3.5 Correlated Work

When several workers are observed at (or near) the same time and their observations are summed into the same x values (and, hence, p values), then their individual reading cannot be considered to be independent.[10] Even if the workers appear to act independently, their activity is correlated (not independent) within an observation round, simply because the observations occur at the same time of the day. For example, an observation round at the end of the day is likely to find all workers in a cleanup mode.

This correlation can be compensated for by calculating for the kth category the alternate standard deviation

$$SD_B = V_B \qquad (4.7)$$

where

$$V_B = \frac{\displaystyle\sum_{j=1}^{J} [Y(k,j)]^2/m(j) - Np^2k}{N(J - 1)} \qquad (4.8)$$

The following definitions describe the quantities in this equation:

J   = total number of observation rounds made during a study.

m(j)   = total number of workers (or machines) observed on the jth observation round of the study. (For example, m(5) is the number of workers observed on the fifth observation round of the study. Ideally, this quantity should be constant throughout a study, but, in practice, it is rarely the case.)

Y(k,j)   = total number of workers (or machines) found in the kth work category on the jth observation round. (For example, Y (3,15) is the number of workers found in Category 3 on the 15th observation round of a study. If only 1 worker is being observed, Y will only be 0 or 1 for each k and j.)

Symbols N and p were defined earlier.

The variance estimator V_B of this section is essentially a scaled sample variance of the Y(k,j) values. Its calculation is greatly simplified with a computer, but can be illustrated with the data in the next example, where the category ''idle'' is taken to be Category 1.

Table 4.2 Given Data Values and Relevant Calculations

Round	No. of workers observed m (j)	No. of workers observed Y (1,j)	$\frac{Y(I,j)^2}{m(j)}$
1	10	7	4.900
2	10	4	1.600
3	14	5	1.786
4	14	2	0.286
5	14	3	0.643
6	16	0	0.000
7	16	4	1.000
8	16	1	0.063
9	16	1	0.063
10	16	2	0.250
11	16	3	0.563
12	16	4	1.000
13	16	0	0.000
14	16	1	0.603
15	16	4	1.000
16	10	0	0.000
Total	232	41	13.217

Example 4.3

Calculate SD_B for the data given in Table 4.2. When, as in the example, only 16 observation rounds are made in a total study, it is appropriate to compensate for the small number by using a percentage point from the t distribution instead of the Z distribution. In practice, however, more than 30 rounds of a study are to be made, and the difference between a t value and a Z value is trivial.

Thus, from Table 4.2 we have

$$P_1 = 41/232 = 0.177$$

$$N = 232$$

and

$$J = 16$$

Substituting the above values into Equation 4.8 yields

$$V_B = [13.217 - 232(0.177)^2]/[232(16 - 1)] = 0.001709$$

Thus,

$$SD_B = \sqrt{V_B} = 0.0413$$

**Table 4.3 p Values for Samples of 50
Observations Each**

		Proportion of Unsafe Acts			
Day	p	Day	p	Day	p
1	0.16	6	0.16	11	0.15
2	0.14	7	0.18	12	0.16
3	0.13	8	0.15	13	0.18
4	0.17	9	0.15	14	0.17
5	0.17	10	0.14	15	0.45

4.3.6   Safety Behavior Control Chart

For developing a control chart, the mean of p (fraction of time each worker is involved in unsafe acts or the mean percent unsafe behavior of the entire group during the observation period) is computed. Using $\bar{p}$, the upper control limit (UCL) and the lower control limit (LCL) can be computed with the aid of the following expressions:

$$\text{UCL} = \bar{p} + 2\sqrt{\bar{p}(1 - \bar{p})/N} \qquad (4.9)$$

$$\text{LCL} = \bar{p} - 2\sqrt{\bar{p}(1 - \bar{p})/N} \qquad (4.10)$$

where $\bar{p}$ is the mean of the p's.
We have used 2 for providing a 95% level* of confidence.

Example 4.4
Table 4.3 shows the results of a sampling behavior study for 50 observations per day. This table shows proportions of unsafe acts observed each day. Using the data of the first 10 days, obtain the values of UCL and LCL of a safety behavior control chart using Equations 4.9 and 4.10. Thus, mean $\bar{p}$ value for the first 10 days = 0.155

Using the above resulting value and N = 50 in Equations 4.9 and 4.10, we get

$$\text{UCL} = 0.155 + 2\sqrt{0.155(1 - 0.155)/50}$$

$$= 0.155 + 0.10 = 0.257$$

$$\text{LCL} - 0.155 - 2\sqrt{0.155(1 - 0.155)/50}$$

$$= 0.155 - 0.10 = 0.055$$

---

* If the value of the LCL is negative, then use zero as the LCL.

This shows that the values for p as given in Table 4.3 for the first 10 days are well within UCL and LCL.

4.3.7 Improving Safety Behavior

In order to improve safety behavior of workers, a major program must be introduced. This could be comprised of safety training programs, lecture series, etc. The safety behavior sampling study may be conducted on a weekly basis during and after the completion of the program. The safety behavior control chart for each period following the start of the program will show if a significant improvement in unsafe behavior has been achieved. Modification of the program or components of the program may be carried out as long as the unsafe behavior is being reduced. Once the minimum of unsafe behavior has been achieved (i.e., p), the behavior sampling study may be repeated and the data plotted on the control chart to assure that unsafe behavior remains at the desired minimum level.

Further use of the data collected can be made in identifying major types of unsafe acts, workers responsible for most of the said unsafe acts, and the time of the day effect on unsafe acts using Pareto's principle [it simply means identify those small percentage (say, 20%), of items for corrective measures which will produce most of the required results (e.g., 80%)]. For more information on Pareto's principle, the reader should consult Reference 11.

4.4 PROBLEMS

1. Describe the following two terms:
    A. Work sampling
    B. Safety behavior sampling
2. Discuss the important factors leading to errors in human performance.
3. Develop expressions for mean and standard deviation of the binomial distribution.

4. Describe the steps involved in the procedure for safety behavior sampling.
5. What is a safety behavior control chart?
6. Discuss the following terms associated with the safety behavior control chart:
   A. Upper control limit (UCL)
   B. Lower control limit (LCL)
7. Construct a control chart for Example 4.4 and discuss the constructed chart.

REFERENCES

1. Johnson, W.G., MORT: *Safety Assurance Systems,* Marcel Dekker, New York, 1980.
2. Heinrich, H.W., *Industrial Accident Prevention,* McGraw-Hill, New York, 1959.
3. Jewel, A.J., An Investigation and Evaluation of the Application of Work Sampling to Safety, Masters Thesis, Ohio State University, Columbus, 1953.
4. Rockwell, T. H., Safety performance measurement, *J. Ind. Eng.* 10, (1), 1959.
5. Schreiber, R., The development of engineering techniques for the evaluation of safety program, *Trans. NY Acad. Sci.,* 18, (3), 1956.
6. Meyer, J.J., Statistical sampling and control for safety, *Ind. Qual. Contr.,* 19, (12), 14, 1963.
7. Pollina, V., Safety Sampling Procedure, prepared by A.O. Smith Corporation, Milwaukee, WI, 1966.
8. Tarrants, W.E., Accident Counsel Factors Obtained from the Critical Incident Study, in *The Measurement of Safety Performance,* Tarrants, W.E., Ed., Garland STPM Press, New York, 1980.
9. Mundel, M.E., *Motion and Time Study: Improving Productivity,* Prentice Hall, Englewood Cliffs, NJ, 1975.
10. Richardson, W.T. and Pape, E.S., *Handbook of Industrial Engineering: Work Sampling,* Wiley, New York, 1984.
11. Dhillon, B.S., *Quality Control, Reliability and Engineering Design,* Marcel Dekker, New York, 1985.

CHAPTER 5

Unsafe Conditions and Contributing Factors in Accidents

5.1 INTRODUCTION

An accident potential exists in almost every situation. Unsafe acts and unsafe conditions coupled with many other factors in the right combination are necessary for an accident to occur. De Reamer[1] has grouped causes of accidents into two categories: (1) immediate causes of accidents, and (2) contributing causes of accidents.

Immediate causes of accidents include unsafe acts and unsafe conditions. On the other hand, contributing causes of accidents include physical conditions of the worker and the management policies. Both groups of these causes are shown in Figure 5.1.

Safety training programs may be used to improve safety behavior of workers, thus minimizing unsafe acts.

A critical examination of task and environmental variables is necessary for improving unsafe conditions. An unsafe condition is defined as any physical condition, which if left uncorrected, is likely to lead to an accident. For improving safety at the work place, such conditions must be detected before an accident occurs.

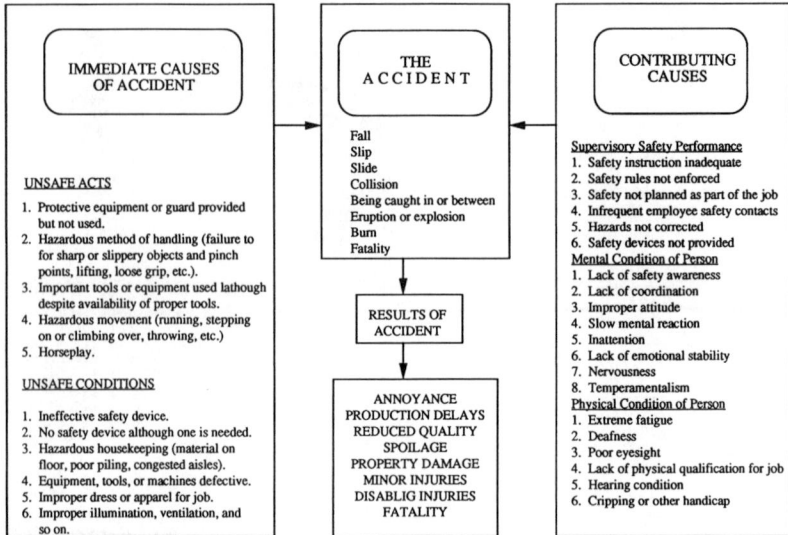

IMMEDIATE CAUSES OF ACCIDENT	THE ACCIDENT	CONTRIBUTING CAUSES

UNSAFE ACTS

1. Protective equipment or guard provided but not used.
2. Hazardous method of handling (failure to for sharp or slippery objects and pinch points, lifting, loose grip, etc.).
3. Important tools or equipment used lathough despite availability of proper tools.
4. Hazardous movement (running, stepping on or climbing over, throwing, etc.)
5. Horseplay.

UNSAFE CONDITIONS

1. Ineffective safety device.
2. No safety device although one is needed.
3. Hazardous housekeeping (material on floor, poor piling, congested aisles).
4. Equipment, tools, or machines defective.
5. Improper dress or apparel for job.
6. Improper illumination, ventilation, and so on.

Fall
Slip
Slide
Collision
Being caught in or between
Eruption or explosion
Burn
Fatality

RESULTS OF ACCIDENT

ANNOYANCE
PRODUCTION DELAYS
REDUCED QUALITY
SPOILAGE
PROPERTY DAMAGE
MINOR INJURIES
DISABLIG INJURIES
FATALITY

Supervisory Safety Performance
1. Safety instruction inadequate
2. Safety rules not enforced
3. Safety not planned as part of the job
4. Infrequent employee safety contacts
5. Hazards not corrected
6. Safety devices not provided
Mental Condition of Person
1. Lack of safety awareness
2. Lack of coordination
3. Improper attitude
4. Slow mental reaction
5. Inattention
6. Lack of emotional stability
7. Nervousness
8. Temperamentalism
Physical Condition of Person
1. Extreme fatigue
2. Deafness
3. Poor eyesight
4. Lack of physical qualification for job
5. Hearing condition
6. Cripping or other handicap

Figure 5.1. Causes of accidents.

5.2 DETECTION OF UNSAFE CONDITIONS

Unsafe conditions can be detected by performing periodic safety inspections. Some of the major features causing unsafe conditions are shown in Figure 5.2.

Formal periodic inspections are usually conducted by plant safety personnel. Numerous checklists are available which may be used for such inspections. One such checklist is presented in Appendix C. Note that the safety inspections may be informal. An informal inspection is usually carried out by a supervisor who ensures that the facilities and equipment are in proper condition at the beginning of the shift. Similar informal inspections are helpful if performed by higher-level supervisors, and the noted discrepancies are brought to the concerned supervisor for taking corrective measures.

5.2.1 Formal Safety Inspections

These inspections are usually conducted by committees comprised of plant safety committees, department managers, fire prevention officials, and plant safety personnel.

For conducting a formal safety inspection, it is necessary that an inspection checklist is used to indicate the conditions of the equipment

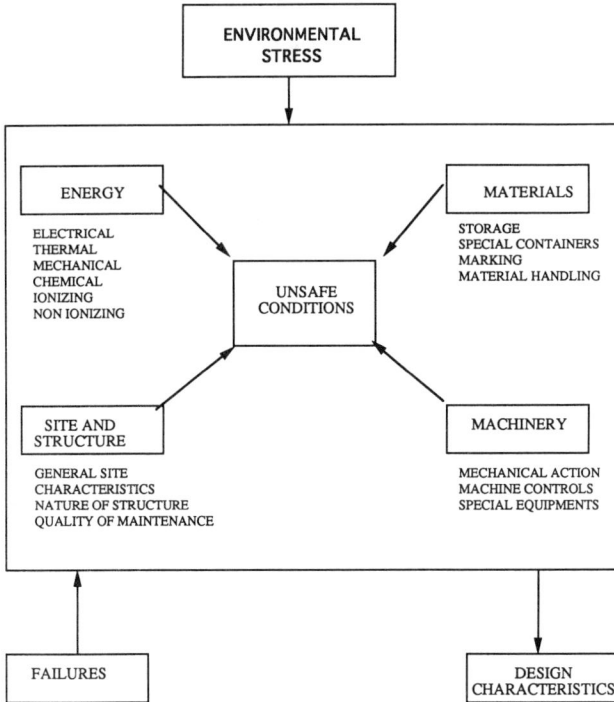

Figure 5.2. Features causing unsafe conditions.

to be inspected and to constitute a record of the findings. One such checklist (with a suggested method of scoring the results of the inspection) is provided in Appendix C.1. The scoring method, as well as the symbols used, are described below.

i = work center classification

X_i = 1, when ith classification is applicable to the plant

X_i = 0, otherwise

S_{ij} = score of the jth item under work center classification i

S_i $= \sum\limits_{S_{ij}}^{n_i}$ total score for work center i

n_i = number of questions under i

$$SS_{i,j} = \sum_{j=1}^{n_i} S_i, j / n_i \qquad (5.1)$$

Table 5.1 Safety Appraisal Scores Compiled by Each Committee Member

Team member	Appraisal 1	R1 rank	Appraisal 2	R2 rank	Appraisal 3	R3 rank
A	78	15	75	12	77	14
B	60	2	64	5	66	7
C	65	6	67	8	71	10
D	57	1	61	3	62	4
E	70	9	72	11	76	13
Total	330	33	339	34	352	48

$$\text{Overall plant score } k = \frac{100}{n} \sum_{i=1}^{n} SS_i/S_i \qquad (5.2)$$

5.2.2 Statistical Inference of Safety Inspections

Every attempt should be made to determine if the inspection scores are showing statistically valid trends. In order to be statistically valid, safety inspection data must be collected over a long period of time. During the data collection period, it is quite possible that the unsafe conditions detected (during the inspections) have been removed; thus, the data based on changed conditions may not be truly representative for determining statistically valid trends. However, the technique described subsequently is suggested to assess the impact of safety inspections on removing unsafe conditions.

Let us assume for a moment that three safety inspections of a certain plant were carried out at three different times by the same safety inspection committee consisting of five members. Table 5.1 presents the totals of safety appraisal scores as compiled by each committee member.

In this case, we may be interested to find out if the total scores of all the safety appraisals belong to the same distribution (i.e., these are not statistically different from each other). In order to achieve this, we can use the Kruskal-Wallis Test described by Bowen and Starr[2] and Raouf.[3]

This test analyzes data in K independent samples to determine whether or not K populations have identical continuous distributions. The hypotheses can be stated as under:

H_0: the K populations are identical
H_a: the K populations are not identical

The following steps are necessary to analyze the inspection scores using this nonparametric test:

1. Rank all data in order of size, assigning 1 to the smallest number. If two or more numbers are equal, each number is assigned as the average rank of the group. Ranking of the data is also shown in Table 5.1.

2. Compute the rank sum of each appraisal; in this case:

    R1 (rank sum for appraisal 1) $= 33$
    R2 (rank sum for appraisal 2) $= 34$
    R3 (rank sum for appraisal 3) $= 48$

3. Compute total sample size (N):

$$N = n1 + n2 + n3$$

In this example, we have $n1 = n2 = n3 = 5$ (i.e., number of team members).

4. Complete H statistic:

$$H = \frac{12}{N(N + 1)} \left[\frac{R1}{n1} + \frac{R2}{n2} + \frac{R3}{n3} \right] - 3(N + 1) \qquad (5.3)$$

using the given data and the results in Equation 5.3, we get

$$\frac{12}{(15)\,(16)} \left[\frac{332}{5} + \frac{342}{5} + \frac{482}{5} \right] - 3\,(16) = 17.84$$

5. Develop the decision rule: reject H_0 as sample H is greater than $\chi^2_{\alpha,\nu}$ value as obtained from χ^{2}* tables

    For $\alpha = .05$ and $\nu = 2$, we get $\chi^2_{.05,2} = 5.991$
    For $\alpha = 0.05$ and $\nu = 1$, we get $\chi^2_{.05,1} = 3.841$

Since Sample $H = 17.84$, which is greater than the critical $\chi^2_{.05,2}$ value of 5.991, we reject the H_0 (i.e., we can say with 95% confidence that each safety appraisal score is statistically different from the other).

Based on the total scores of each appraisal, we can say that unsafe conditions are showing the trends of improvement.

---

* It is popular distribution used for hypothesis testing and is a special case of the gamma distribution.[4]

The score sheets are indicative of the unsafe conditions in the plant and can be used to constitute a record of findings. Findings during previous inspections are to be reviewed to determine if the discrepancies have been corrected. The score sheets should invariably present recommendations for corrective actions, and this procedure, if followed, can provide needed information to safety personnel for evaluating the results of a safety improvement program.

The result of safety inspections can be analyzed with a view to identify areas more hazardous than others. It is a common practice to color code a map of the plant as high hazard, moderate hazard, and low hazard areas. Safety inspections are to be conducted with varying frequencies of visits to areas with different degrees of hazard. The most frequent inspection should be carried out at high hazard areas and the next priority should be given to moderate hazard areas. Low hazard areas should be inspected less frequently than the other two areas.

5.3   CONTRIBUTING CAUSES OF ACCIDENTS

To completely understand the causes of an accident, a full understanding of the contributing and immediate causes of the accident is essential. In the past, unsafe acts and unsafe conditions were emphasized for preventing the occurrence of accidents. Contributing causes of an accident, if corrected before the accident happens, are most likely to result in accelerating the accident prevention process.

5.3.1   Supervisor's Safety Performance

The safety performance of a supervisor may be evaluated by examining the following factors:

Job Hazard Analysis
The supervisor is required to perform hazard analysis of the jobs carried out in the high and medium hazard areas of the plant at minimum. This analysis can be conducted on the same basis as the production planning (i.e., specifying tools, machines, inspection methods, etc.).

Enforcement of Safety Rules
It is the responsibility of the safety supervisor to ensure that safety rules are applied equally to all with fair and square attitude.

Adequate Safety Knowledge

It is the responsibility of the supervisor to impart necessary safety knowledge to his workers. This is usually accomplished through training.

Workers Participation in Safety

Quality circles[5] concept can be used in improving safety. This approach is based on the principle of forming a voluntary team to study the methods of improving safety.

Proper Job Placement

Supervisors should try to match workers capabilities to task requirements. A mismatch is likely to result in unsafe acts performed by the workers.

Improving Safe Working Conditions

The safety inspection results must be considered with a view to incorporate the suggestions made by the inspection team.

5.3.2 Mental Condition of Workers

The mental condition of workers is an important factor with respect to safety. It is useful to consider the following factors.

Safety Awareness

This calls for supervisors to improve their workers' safety awareness by offering regular safety courses.

Eye-Hand Coordination

Lack of eye-hand coordination can result in accidents. This can be improved by providing necessary training to workers. The eye-hand coordination becomes a factor that is even more imperative if workers with required capabilities are not available, especially in the case where government regulation prohibits management from discriminating workers.

Workers' Attitude

Improper attitude of the workers may cause accidents. This may be rectified by safety promotion and publicity to some degree.

Reaction and Emotional Stability

If the worker is slow in reacting to stimuli and is not emotionally stable, then he or she is likely to be involved in accidents.

5.3.3 Physical Condition of Workers

This is another important factor which calls for careful attention to the following:

1. To find a worker with the required physical condition for performing his/her task, preplacement medical examinations are necessary.
2. Periodic reexaminations are needed to ensure that the environmental exposure is not affecting the worker's physical condition adversely.
3. In the event a worker is transferred from one job to another, a careful check of his/her physical conditions should be conducted.

5.3.4 Evaluation of Contributing Conditions for Accidents

Based on the above information and the factors identified in Figure 5.1, a checklist can be developed. Safety inspection teams can include this evaluation in their program. Results of periodic inspections can be tested for statistical trends using the method outlined in Section 5.2.2. A suggested scoring sheet is given in Appendix C.2.

5.4 DEVELOPING A COMPOSITE SCORE FOR SAFETY PERFORMANCE

As the quantitative assessment of safety performance is an area of great importance, we present two methodologies to develop a composite score for the performance in this section. Both methodologies may be used to monitor a safety program.

Petersen[6] has suggested a scheme for measuring Industrial Safety Activities based on Dickemper and Spartz's work.[7] In this scheme, activities are grouped into the following five categories:

1. organization and administration
2. industrial hazard control

Table 5.2 Weights for Five Groups of Activities

No.	Category description	Weight (%)
1	Organization and administration	30
2	Industrial hazard control	25
3	Fire control and industrial hygiene	20
4	Supervisor participation, motivation, and training	15
5	Accident prevention and reporting procedures	10

3. fire control and industrial hygiene
4. supervisor participation, motivation, and training
5. accident investigation and reporting procedures

Each of the above categories is further broken down into several activities. The score for each category is between Poor (0), Fair (26), Good (80), and Excellent (100). The weight of each category for determining the overall score is fixed at 20%. This aspect of the scheme was discussed by Petersen[6] with a group of senior safety professionals during a refresher course in Saudi Arabia.

The DELPHI Method[8] scheme was used to determine the weight of each of the above categories. Table 5.2 presents weights determined for each of the above five categories.

The second composite score-developing scheme for safety performance calls for the evaluation of the following factors:

1. Safety Behavior Sampling (SBS)
2. Unsafe Conditions (USC)
3. Contributing Factors (CFS)

These factors are to be monitored individually and their individual scores determined out of a total of 100. In order to develop a composite score of safety program, the following suggestions are useful:

1. Assign weight to each factor.
2. Sum up weighted scores.

The composite score (CS) is given by

$$CS = K_1(SBS) + K_2(USC) + K_3(CFS) \qquad (5.4)$$

where K_1, K_2, and K_3 are the weights for corresponding factors (i.e., SBS, USC, and CFS, respectively). Their values may be taken as 0.5, 0.3, and 0.2, respectively.

Table 5.3 Data for Example 5.1

	Appraisal			
Inspection	**1**	**2**	**3**	**4**
Safety Behavior Sampling Score (SBS)	75	60	79	70
Unsafe Conditions Score (USC)	80	70	95	60
Contributing Factor Score (CFS)	50	60	75	55
Composite Score (CS)	71.5	63.0	83.0	64.0

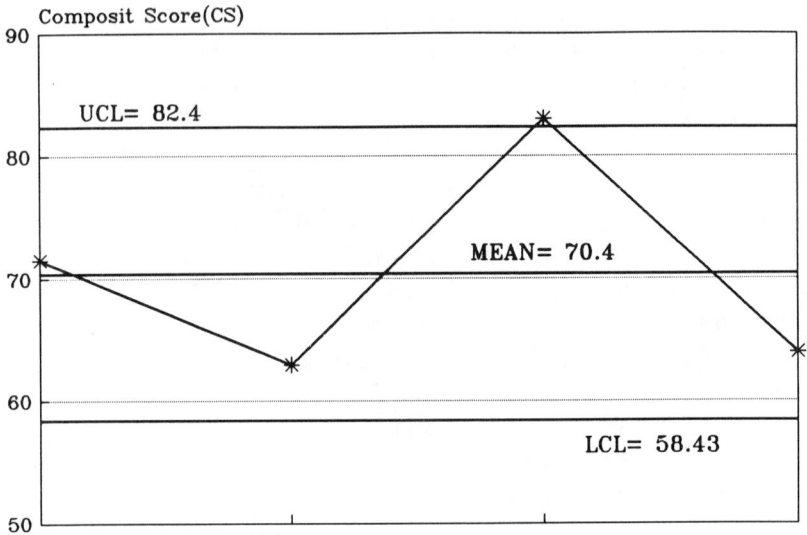

Figure 5.3. Composite Score (SC) Control Chart. UCL = upper control limit; LCL = lower control limit.

The sum of scores so obtained can be treated as a composite safety score. A control chart of CS can be plotted to detect undesired fluctuations. If the CS for a given period falls outside the control limits of the chart, then an investigation must be initiated to locate the assignable causes for this variation.

Example 5.1

This example demonstrates the application of this scheme by using the data given in Table 5.3.

The resulting control chart is shown in Figure 5.3. It shows that the CS score for Appraisal 3 is above the upper control limit (UCL). This calls for an investigation to find out its assignable cause. The term LCL in Figure 5.3 means lower control limit.

5.5 PROBLEMS

1. Discuss the contributing causes of accidents.
2. List immediate causes of accidents.
3. Discuss the factors to be examined in evaluating supervisor's safety performance.
4. Discuss the safety related factors to be considered with respect to the mental condition of workers.
5. Describe a methodology useful to develop a composite score for safety performance.

REFERENCES

1. De Reamer, R., *Modern Safety and Health Technology,* John Wiley & Sons, New York, 1980.
2. Bowen, K.E. and Starr, M.K., *Basic Statistics for Business and Economics,* McGraw-Hill, New York, 1982.
3. Raouf, A., Conducting Safety Appraisals and Drawing Statistical Inference, Hazard Prevention, 4th Quarter, 1989.
4. Tsokos, C.P., *Probability Distributions: An Introduction to Probability Theory with Applications,* Wadsworth Publishing Company, Belmont, CA, 1972.
5. Ishikawa, K., *Quality Control Circles at Work: Cases from Japan's Manufacturing and Service Sectors,* Asian Productivity Organization, Tokyo, 1984.
6. Petersen, D., *Human Error Reduction and Safety Management,* Garland STPM Press, New York, 1982.
7. Dickemper, R. and Spartz, Z., A quantitative and qualitative measurement of industrial safety activities, *ASSE J. Am. Soc. Saf. Eng.,* Dec., 1970.
8. Dalkey, N. and Helmers, O., An experimental application of the Delphi method to the use of experts, *Manage. Sci.* 9, (3), 1963.

CHAPTER 6

Safety Information System

6.1 INTRODUCTION

Safety professionals spend a significant amount of time sorting through various files to retrieve useful data. In some cases, they are required to post a log of occupational injuries and illnesses by a certain date at a specified place in the plant, e.g., Occupational Safety and Health Act (OSHA) Form No. 200. Some companies may require the collection of Accident Statistics to define those areas that are in need of attention, the order of priority, and the type and size of resources required to implement corrective measures. Before such actions can be taken, the available data have to be transformed into information. Typical sources of data are given in Figure 6.1.

An example of monthly accident statistics compiled by an organization having four departments (A, B, C, D) is shown in Table 6.1. Table 6.2 presents descriptions of columns 1 through 13 and of month and year.

A computerized Safety Information System (SIS) enables a safety manager and others to prepare reports to meet legal obligations of the company and to initiate appropriate corrective measures. This helps

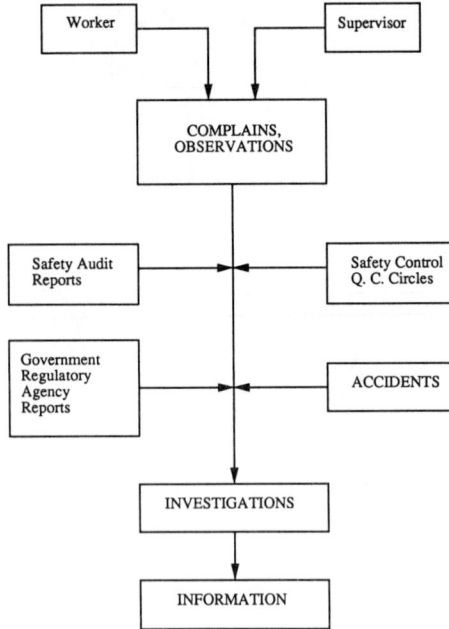

Figure 6.1. Sources of safety related information.

Table 6.1 Accident Statistics Compiled by an Organization for Month 12 — Year 1987

1	2	3	4	5	6	7	8	9	10	11	12	13
A	30	2800	10	2083	2	10	417	1250	174	521	6	30
B	25	4000	5	1250	1	5	250	750	104	313	3	15
C	28	4480	4	893	1	5	223	670	93	298	3	16
D	28	5000	4	800	1	5	200	600	83	267	3	16
M	111	11280	23	1258	5	—	274	821	—	—	15	—
Year	—	219360	240	1094	—	25	—	—	114	351	—	77

to reduce manpower requirements while increasing the productivity of the safety group.

This chapter describes the SIS computer package utilizing data base management system package dBase III. Details of this package and a user manual are provided in Appendices A, D.1, and D.2. Some of the salient features of the package are outlined in the sections that follow.

---

* The dBase-compatible computer program can be obtained free of charge from the authors, whose addresses are in the front of the book.

Table 6.2 Descriptions of Column 1 through 13 and of Month and Year

Column (plus month and year)	Description
1	Departments of companies or industries name (plus M and year)
2	Number of workers employed
3	Man-hours worked in current month
4	Number of reported injuries for the current month
5	Frequency rate of injuries for the month
6	Number of Loss Time Accidents for the current month (LTAs)
7	Number of cumulative LTAs
8	Frequency rate of LTAs for current month
9	Severity rate of LTAs for current month
10	Cumulative frequency rate of LTAs
11	Cumulative severity rate of LTAs
12	Number of days lost in current month
13	Cumulative number of days lost
M	Total monthly figures
Year	Cumulative figures up to end of current month

6.2 SAFETY INFORMATION SYSTEM (SIS)

The purpose of SIS is threefold: (1) store relevant accident and investigation data, (2) process stored data, and (3) present final results in report form for assisting safety managers in assessing their safety performance and helping them in initiating necessary corrective actions. SIS allows a continuous updating of pertinent data, display of this information, and the capability of performing measurements and evaluations of safety performance.

6.2.1 Data Base

The items included in the data base are shown in Table 6.3. The details of coding such data are given in Appendix D.2.

6.2.2 Summary of the Investigation

SIS possesses the capability to process not only the accident investigations, but investigations of near accidents as well, in addition to two classes of investigations: pre-accident investigations and post-accident investigations. The pre-accident investigations can be the result

Table 6.3 Data Base Items

Item	Description
1	Case of file number
2	Year
3	Month
4	Day
5	Investigation type
6	Employee department code
7	Department code where accident occurred
8	Location code
9	Occupation code
10	Worker number
11	Age group
12	Sex
13	Experience group
14	Nationality code
15	Day of week
16	Hour
17	Activity code
18	Source of injury
19	Accident type
20	Hazardous condition
21	Unsafe act
22	Nature of injury
23	Part of body
24	Hours worked since last day off job
25	Overtime worked since last day off job
26	Severity: lost work days
27	Severity: total cost in riyals
28	Investigator code

of formal inspections that were randomly conducted by safety personnel. Furthermore, special investigations may be generated upon the request of the worker or the safety supervisor.

Post-accident investigations are carried out to investigate an accident. One must note here that the recordable injuries (in terms of OSHA) are not the only source of post-accident investigations. Accidents that result in no injuries also provide important information regarding potential hazards. SIS can provide a summary of all the accident statistics presented in Table 6.1.

6.2.3 Statistical Analysis of Accident Statistics

This program provides univariate as well as bivariate distributions and carries out statistical comparisons.

The capability of SIS to sort data through many items, or combination of items, makes it possible to generate reports that can identify the correlation that may exist between any variable(s) and the frequency

Table 6.4 Variable List

Variable descriptions
Year
Month
Day
Investigation type
Employee department
Accident in department
Location
Occupations
Age group
Sex
Experience group
Nationality
Day of the week
Time in hours
Activity
Source of injuries
Type of accident
Hazardous condition
Hours worked since last day off
Overtime hours worked since last day off

of accidents. Such information goes a long way in providing insight into the causes of accidents.

There are various types of reports that may be considered helpful to safety managers for the purpose of measuring, evaluating, and increasing control over safety performance. A set of reports were selected in an attempt to give an effective feedback of the safety performances without providing redundant information.

6.2.3.1 Univariate Distribution

For all the variables listed in Table 6.4, a frequency of occurrence, lost days, and total cost associated with the variable in question value, if any, are provided. This gives a great advantage to the user since it indicates potential causes of accidents and their contributions to cost. For example, when age group is the variable, it becomes very helpful in observing the relationship that exists between the age-groups and the frequencies of accidents. Furthermore, when a specific age-group(s) appears to cause more accidents, certain actions may be taken to reduce accidents. The user can generate up to 20 different univariate distribution reports.

A list of all variables to be used in univariate distribution is given in Table 6.4. These variables will also be used for bivariate distribution.

6.2.3.2   Univariate Distribution by Department

This report is the same as the univariate distribution report in principle, but the only difference is that it does not show the lost days and cost. The report lists the variable values for all departments, which will serve as a means of making comparisons among the departments. The departments represented here are the units where accidents occur.

6.2.3.3   Bivariate Distribution

In this case, the same principle used for the construction of a univariate distribution is applied to construct a bivariate distribution. This report tabulates two variables simultaneously, thus enabling the user to recognize additional correlations between variables. For example, consider age-group and the cause of accident. A bivariate distribution of these two variables will indicate any relationship that exists between the age of the worker and the cause of the accident. This can help the user when the general age distribution does not vary from the age distribution of those involved in accidents, and the same is applied for the causes of accidents. Any two variables in Table 6.4 can be used to produce a bivariate distribution.

6.2.3.4   Statistical Calculation

This is not an actual report but rather a requirement for conducting the statistical comparisons. The statistical calculations, however, may be used as a report to show the user the average number of recordable injuries per month for a certain historical data and the standard deviation. Those values are stored in the system for each department for later reference by the statistical comparison report. All in all, it may be said that they serve as a reference to calculate the behavior of injury occurrence for a subsequent period of time.

6.2.3.5   Incidence Rates

The incidence rate is a statistical report used to provide a statistical basis to describe the number of events that are occurring in a given period of time.

This system uses 2,000,000 man-hours as a base and includes injury and the lost days cases. These cases are used in turn to calculate the following indices:

$$\text{Injury Rate} = \frac{\text{No. recordable injury cases} \times 2,000,000}{\text{workers-hours}} \quad (6.1)$$

$$\text{Frequency Rate} = \frac{\text{No. lost days cases} \times 2,000,000}{\text{workers-hours}} \quad (6.2)$$

$$\text{Severity Rate} = \frac{\text{Total lost days} \times 2,000,000}{\text{workers-hours}} \quad (6.3)$$

The worker-hours are approximated by:

$$\begin{aligned}
\text{worker-hours} = &\ (\text{plant population average per month}) \\
&\times (\text{no. work-days in the month}) \quad (6.4) \\
&\times (\text{no. shifts}) \times (\text{duration of shift})
\end{aligned}$$

6.2.3.6 Statistical Comparison

This report utilizes the test of hypothesis to compare the safety system performance in a given period of time with the historical data measures to determine whether the safety system performance has undergone significant changes.

For the comparison purpose, the monthly injury frequencies are used. For example, one can compare the monthly injury frequencies of the previous 6 months with the last 4 years to observe the effect of implementing a new safety program. This report makes the comparison of more than 5 months since the test of hypothesis results may not be values for small n (i.e., less than 6).

A two-tailed test of hypothesis is implemented using the t-distribution. In addition, 0.01 and 0.05 levels of significance are used in performing the comparison.

6.2.3.7 Data Listing

The SIS stores information that can readily be accessed and retrieved. The information is displayed in the form of codes and descriptions. The system enables the user to observe accidents or investigations that have some similarities within a given span of time. This provides the user the opportunity of examining cases that have a common factor or factors among them and obtaining the relations between them (if any).

6.2.3.8   Univariate Listing

This listing utility allows the user to list investigations that occurred in a certain period in a given department, according to some common variable(s). For example, the user can look up the accidents occurring in January 1986 in the Warehouse Department committed by the workers of the Shipping and Receiving Department. The variables allowed here are the same variables used by the univariate distribution report.

6.2.3.9   Univariate Listing by Department

This listing utility allows the user the same selection facilities allowed in univariate listings, but the accidents are listed in respect to all departments.

6.2.3.10   Bivariate Listing

The bivariate listing allows the user to retrieve investigations with respect to two variable values. For example, the user may like to look up the accidents that occur on, say, holidays.

6.3   CONCLUDING REMARKS

This chapter described SIS, which can be used to improve safety performance. SIS is designed to help safety professionals measure and evaluate their safety performance and indicate areas for improvement.

6.4   PROBLEMS

1. What is the difference between the univariate and bivariate distributions?
2. Write an essay on the Safety Information System (SIS).
3. What are the sources of safety related information?
4. Define the following incidence rates:
    A. Injury rate
    B. Frequency rate

---

*   The dBase-compatible computer program can be obtained free of charge from the authors, whose addresses are in the front of the book.

CHAPTER 7

Automation and Safety

7.1 INTRODUCTION

Automation is defined as the technology concerned with the application of complex mechanical, electronic- and computer-based systems in the operation and control of production systems.[1] Reference 2 states that automation should be viewed as a trend and not as a state associated with the replacement of human activities by machines. Most companies are trying to automate their production operations with a view to improve productivity and quality of their products. Such companies are adopting automation in a progressive manner. An approach frequently used is to form "islands of automation" and ultimately bridge such islands to obtain an integrated and automated factory.[3]

7.2 THE EFFECT OF AUTOMATION ON THE WORKER'S PERFORMANCE

It is generally agreed that automation affects direct and indirect manual work, knowledge and skill requirements of workers, social

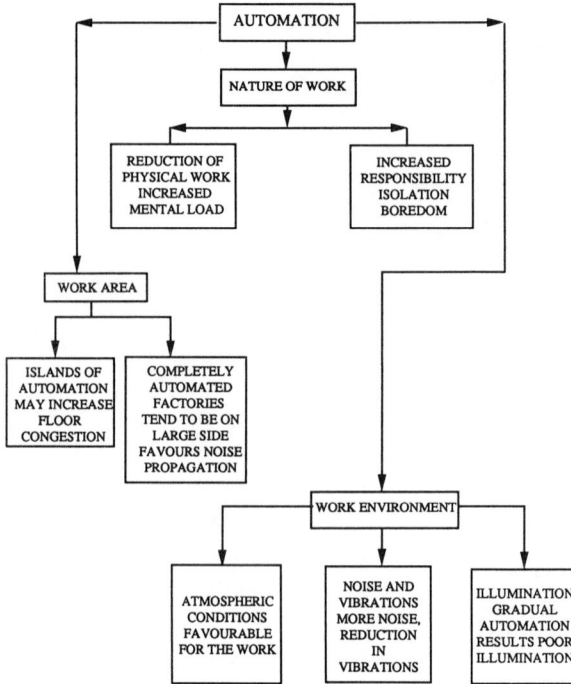

Figure 7.1. Affect of automation on work-related factors.

interaction among workers, pace of work, physical effort requirements, and the working environment.

In general, automation results in a reduction of the number of workers per unit floor area of the plant, and in some cases, it separates workers from machines and increases mental load on the worker while reducing physical effort on his or her part. Figure 7.1 shows some of the salient factors directly influenced by automation.

As stated earlier, automation is to be viewed as a "trend" and not as a "state". Its effects on the factors mentioned are going to be a function of "automation level" at a given time. It may be said that for a given level of automation, there could be a certain level of homogeneity among factors given in Figure 7.1, thus nullifying some of the shortcomings of standard measures of safety performance.[2]

A method of scaling the level of automation is presented below. It allows comparisons of safety performance among departments, plants, etc. having similar levels of automation.

7.2.1 Scaling Automation Level

A typical manufacturing operation may be broken down into three components as follows:

1. Processes
2. Material Handling Systems
3. Controls

Processes include machining, assembling, forming, and so on, and material handling systems comprised of transportations, storage, load or unload activities, etc. Components of controls include items such as activation of machines and monitoring of operations. The level of automation (LA) is expressed as a function of processes, material handling systems, and control automation as follows:

$$LA = f(AP_i, AM_j, AC_k) \qquad (7.1)$$

where AP_i = ith level of process automation
$\quad\quad\;\; AM_j$ = jth level of material handling system automation
$\quad\quad\;\; AC_k$ = kth level of control automation

The means of dimensioning LA are given in References 3 and 4. Using such dimensions, a manufacturing plant can be categorized into the following distinct levels:

AP_i . . . i = 1,2,3,4,5
$\quad\quad\quad\quad\;$ 1 = hand tools used for carrying out operations
$\quad\quad\quad\quad\;$ 2 = powered hand tools
$\quad\quad\quad\quad\;$ 3 = hand-controlled machine
$\quad\quad\quad\quad\;$ 4 = automatic machine cycle after operator activates it
$\quad\quad\quad\quad\;$ 5 = completely automatic machining operation

AM_j . . . j = 1,2,3,4,5
$\quad\quad\quad\quad\;$ 1 = material handled manually
$\quad\quad\quad\quad\;$ 2 = powered tools used for material handling
$\quad\quad\quad\quad\;$ 3 = hand-controlled machines handling material
$\quad\quad\quad\quad\;$ 4 = automatic machines handling material after operator
$\quad\quad\quad\quad\quad\quad\;$ initiates it
$\quad\quad\quad\quad\;$ 5 = completely automatic machines handling material

AC_k . . . $k = 1,2,3,4,5$

         1 = worker performs all control functions manually

         2 = machine started by a timer or a product; remainder of control functions performed manually

         3 = machine activation and monitoring automated; remainder of functions performed manually

         4 = automatic machine activation, process monitoring, and adjustment; other control functions performed manually

         5 = fully automatic

7.2.2 Relationship Between Level of Automation and Work Requirements

Different LAs affect the worker's requirement and performance. Each of these is briefly described below.

Direct Manual Work (DMW) — DMW is maximum when LA is minimum, and it decreases as LA increases. DMW is zero when LA is maximum.

Indirect Manual Work (IMW) — IMW is nonexistent when LA is minimum and is maximum when LA is maximum.

Knowledge and Skill Requirements (KSR) — When LA is minimum, KSR is the highest. This may reduce to zero when LA is midway between maximum and minimum and may start increasing beyond LA midpoint. This essentially is due to maintenance and repair work needed for higher LAs.

Social Interaction (SI) — SI is maximum when LA is minimum. Since increase in LA means moving the worker away from the worksite, SI decreases when LA increases. SI may increase if many operatives are housed in one control room.

Work Pace (WP) — WP is worker-controlled for lower values of LA, but worker loses control over his pace of work as LA increases.

Physical Effort (PE) — For lower levels of LA, PE is maximum, and it nearly diminishes when LA reaches maximum.

Working Conditions (WC) — Working conditions improve significantly as LA increases.

7.3 IMPROVING SAFETY IN AUTOMATED PLANTS

Development in manufacturing technologies results in the change of the level of automation. These developments are in the area of automated production equipment and automated materials handling. The changes occurring in these areas are creating unrecognized hazards.

It may be said that no matter how automated a plant may become, there will always be a human element of some sort involved in its operation. Some of the functions* typically performed by workers in automated systems are listed:

1. removing material
2. inserting parts
3. program creation/modification
4. adjusting tools, clamping devices
5. monitoring the system
6. locating and fixing malfunctions
7. services and maintenance

In the following, a brief description of sources of automation-related hazards is presented. This should assist the safety professional in developing a list of "unsafe conditions" and "unsafe acts" pertinent to his/her organization. Subsequently, using the techniques of safety behavior sampling and safety appraisals, the safety manager can analyze, predict, and evaluate safety performance.

7.3.1 Sources of Automation-Related Hazards

Automation-related hazards may be classified into the following five categories.

7.3.1.1 Control

Control errors may be caused by faults within the control systems, errors in software, or electrical interferences. Reference 5 reports that 83% of the accidents were due to inadvertent start, 5% failure to stop, 11% intended movements, and 1% abnormal movement.

7.3.1.2 Presence in the Working Envelop of the Machine

When the machine is operating in the automatic mode, the operator's presence in the work envelop of the machine is normally prevented

---

* Presence of these tasks depends on the level of automation of the system.

Table 7.1 Breakdowns of the Human Error with a Control System

No.	Type of error	Number
1	Unintentional contact with start devices	29
2	Unintentional contact with other switch(es)	17
3	Erroneous movement	8
4	In repetitive work	—
5	Mistakes of maneuver	14
6	Unknown function to the operator	14
7	Did not know other person was in danger zone	12
8	Unclear	4
	Total	98

by suitable safety devices. The operator may enter the work envelop of the machine due to reasons such as:

1. The machine is not properly guarded.
2. Some manual operations are necessary.
3. The operator is required to perform repairs while the machine is running, i.e., fault location, carrying out necessary adjustments.
4. The design of the equipment is such that it is not possible to turn off the power.
5. The programmer is using "teach-in" method, i.e., the axis are moved by means of a manual programming device and, thus, the movements created are stored.

More information on this topic is given in Reference 5.

7.3.1.3 Human Error

The automated equipment differs drastically from the equipment it replaces and requires sophisticated maintenance and operation skills. If operators are not adequately trained, it is difficult to anticipate how they will react in emergency situations.

Reference 6 reports that in 98 cases, a human error with a control system was identifiable; the breakdowns are given in Table 7.1.

Programmers usually work in close spaces for long periods of time, and their working postures are generally undesirable. The lack of adequate light, coupled with other factors, can cause additional stress on humans.

Reference 7 reports data on 131 cases and presents the degree of risk for various type of workers as given in Table 7.2.

Table 7.2 Degree of Risk for Various Types of Workers

No.	Worker type	Frequency (%)
1	Programmers	57
2	Trouble shooters	26
3	Operating personnel (normal operation)	13
4	Maintenance	4
	Total	100

The risk to programmers may be minimized by observing the following:

1. Keep the robot area of movement as small as possible.
2. Plan programmers' location outside the area of robot movement.
3. Minimize postural and environmental stresses.

7.3.1.4 Electrical, Hydraulic, and Pneumatic Faults

In this case, depending on the type of machine control device, the chances of fire, high pressure leak, etc. still exist.

7.3.1.5 Mechanical Hazards

These may result from parts or materials carried or by overloading or by degradation of the reliability of the system due to corrosion or fatigue.

7.4 AUTOMATED EQUIPMENT SAFETY

Many safety requirements are placed on automated equipment by national codes and laws. In addition to mandatory government codes, there are other standards set by organizations whose requirements are often important. Some of those organizations are:

1. Occupational Safety and Health Administration (OSHA)
2. American National Standards Institute (ANSI)
3. Deutsche Institute Fuer Normung (DIN)
4. Health and Safety Executives (HSE)

All in all, it can be said that before installing automated equipment, ensure that all the necessary safety-related requirements are fulfilled.

7.5 SOME SUGGESTIONS FOR IMPROVING SAFETY IN AUTOMATED PLANTS

Generally, at present, rigid rules are not available for handling safety at automated plants. However, each hazard and risk of injury must be evaluated, and safety procedures analyzed. Based on this information, no matter how little it may be, techniques like safety-behavior sampling and safety appraisals, coupled with trend analysis, can be helpful to analyze, predict, and evaluate safety performance of automated plants. The collected information can also be used for the purpose of developing data banks and expert systems.

Last, but not least, appropriate worker selection procedures and training programs must be followed, and their success should be evaluated periodically.

7.6 PROBLEMS

1. Discuss the effects of automation on a worker's performance.
2. List the number of functions typically performed by workers in automated systems.
3. Discuss the sources of automation-related hazards.
4. What are the possible reasons for a human entering the work envelop of an automated machine in a plant?
5. Define human error.
6. Discuss the possible ways to reduce risk to automated system programmers.

REFERENCES

1. Groover, M., *Automatic Production Systems,* Computer Integrated Manufacturing, Wiley, New York, 1987.
2. Raouf, A., Effects of automation on occupational safety and health, in *Ergonomics of Hybrid Automated Systems,* Karwowski, W., Ed., Elsevier, Amsterdam, 1988.
3. Raouf, A., Automation — its measurement and effects on working conditions, in *Trends in Ergonomic/Human Factors III,* Karwowski, W., Ed., Elsevier, Amsterdam, 1986.
4. Wild, R., *Production and Operations Management,* Holt, Rinehart, & Winston, New York, 1979.
5. Dhillon, B.S., *Robot Reliability and Safety,* Springer-Verlag, New York, 1991.
6. Backstop, T. and Harms, R., A statistical study of control systems and accidents at work, in *Occupational Accident Research,* Kjellen, U., Ed., Elsevier, Amsterdam, 1984.
7. Nicolaisen, P., Ways of improving industrial safety for the programming of industrial robots, Proc. 3rd Int. Conf. Human Factors in Manufacturing, Stratford-upon-Avon, U.K., 1986.

APPENDIX A. CHAPTER 3 TABLES

Table A.1 The Poisson Distribution

$$P(X \le x) = \sum_{\kappa=0}^{x} \frac{\lambda^k e^{-\lambda}}{k!}$$

					$\lambda = E(X)$					
x	**0.1**	**0.2**	**0.3**	**0.4**	**0.5**	**0.6**	**0.7**	**0.8**	**0.9**	**1.0**
0	0.905	0.819	0.741	0.670	0.607	0.549	0.497	0.449	0.407	0.368
1	0.995	0.982	0.963	0.938	0.910	0.878	0.844	0.809	0.772	0.736
2	1.000	0.999	0.996	0.992	0.986	0.977	0.966	0.953	0.937	0.920
3	1.000	1.000	1.000	0.999	0.998	0.997	0.994	0.991	0.987	0.981
4	1.000	1.000	1.000	1.000	1.000	1.000	0.999	0.999	0.998	0.996
5	1.000	1.000	1.000	1.000	1.000	1.000	1.000	1.000	1.000	0.999
6	1.000	1.000	1.000	1.000	1.000	1.000	1.000	1.000	1.000	1.000

x	**1.1**	**1.2**	**1.3**	**1.4**	**1.5**	**1.6**	**1.7**	**1.8**	**1.9**	**2.0**
0	0.333	0.301	0.273	0.247	0.223	0.202	0.183	0.165	0.150	0.135
1	0.699	0.663	0.627	0.592	0.558	0.525	0.493	0.463	0.434	0.406
2	0.900	0.879	0.857	0.833	0.809	0.783	0.757	0.731	0.704	0.677
3	0.974	0.966	0.957	0.946	0.934	0.921	0.907	0.891	0.875	0.857
4	0.995	0.992	0.989	0.986	0.981	0.976	0.970	0.964	0.956	0.947
5	0.999	0.998	0.998	0.997	0.996	0.994	0.992	0.990	0.987	0.983
6	1.000	1.000	1.000	0.999	0.999	0.999	0.998	0.997	0.997	0.995
7	1.000	1.000	1.000	1.000	1.000	1.000	1.000	0.999	0.999	0.999
8	1.000	1.000	1.000	1.000	1.000	1.000	1.000	1.000	1.000	1.000

Table A.1 The Poisson Distribution (*continued*)

x	2.2	2.4	2.6	2.8	3.0	3.2	3.4	3.6	3.8	4.0
0	0.111	0.091	0.074	0.061	0.050	0.041	0.033	0.027	0.022	0.018
1	0.355	0.308	0.267	0.231	0.199	0.171	0.147	0.126	0.107	0.092
2	0.623	0.570	0.518	0.469	0.423	0.380	0.340	0.303	0.269	0.238
3	0.819	0.779	0.736	0.692	0.647	0.603	0.558	0.515	0.473	0.433
4	0.928	0.904	0.877	0.848	0.815	0.781	0.744	0.706	0.668	0.629
5	0.975	0.964	0.951	0.935	0.916	0.895	0.871	0.844	0.816	0.785
6	0.993	0.988	0.983	0.976	0.966	0.955	0.942	0.927	0.909	0.889
7	0.998	0.997	0.995	0.992	0.988	0.983	0.977	0.969	0.960	0.949
8	1.000	0.999	0.999	0.998	0.996	0.994	0.992	0.988	0.984	0.979
9	1.000	1.000	1.000	0.999	0.999	0.998	0.997	0.996	0.994	0.992
10	1.000	1.000	1.000	1.000	1.000	1.000	0.999	0.999	0.998	0.997
11	1.000	1.000	1.000	1.000	1.000	1.000	1.000	1.000	0.999	0.999
12	1.000	1.000	1.000	1.000	1.000	1.000	1.000	1.000	1.000	1.000

Reprinted with the permission of Macmillan Publishing Company from *Probability and Statistical Inference*, Third Edition, by Robert V. Hogg and Elliot A. Tanis. Copyright© 1988 by Macmillan Publishing Company.

Table A.2 Values of the Standard Normal Distribution Function

z	0	1	2	3	4	5	6	7	8	9
−3.0	.0013	.0010	.0007	.0005	.0003	.0002	.0002	.0001	.0001	.0001
−2.9	.0019	.0018	.0017	.0017	.0016	.0016	.0015	.0015	.0014	.0014
−2.8	.0026	.0025	.0024	.0023	.0023	.0022	.0021	.0021	.0020	.0019
−2.7	.0035	.0034	.0033	.0032	.0031	.0030	.0029	.0028	.0027	.0026
−2.6	.0047	.0045	.0044	.0043	.0041	.0040	.0039	.0038	.0037	.0036
−2.5	.0062	.0060	.0059	.0057	.0055	.0054	.0052	.0051	.0049	.0048
−2.4	.0082	.0080	.0078	.0075	.0073	.0071	.0069	.0068	.0066	.0064
−2.3	.0107	.0104	.0102	.0099	.0096	.0094	.0091	.0089	.0087	.0084
−2.2	.0139	.0136	.0132	.0129	.0126	.0122	.0119	.0116	.0113	.0110
−2.1	.0179	.0174	.0170	.0166	.0162	.0158	.0154	.0150	.0146	.0143
−2.0	.0228	.0222	.0217	.0212	.0207	.0202	.0197	.0192	.0188	.0183
−1.9	.0287	.0281	.0274	.0268	.0262	.0256	.0250	.0244	.0238	.0233
−1.8	.0359	.0352	.0344	.0336	.0329	.0322	.0314	.0307	.0300	.0294
−1.7	.0446	.0436	.0427	.0418	.0409	.0401	.0392	.0384	.0375	.0367
−1.6	.0548	.0537	.0526	.0516	.0505	.0495	.0485	.0475	.0465	.0455
−1.5	.0668	.0655	.0643	.0630	.0618	.0606	.0594	.0582	.0570	.0559
−1.4	.0808	.0793	.0778	.0764	.0749	.0735	.0722	.0708	.0694	.0681
−1.3	.0968	.0951	.0934	.0918	.0901	.0885	.0869	.0853	.0838	.0823
−1.2	.1151	.1131	.1112	.1093	.1075	.1056	.1038	.1020	.1003	.0985
−1.1	.1357	.1335	.1314	.1292	.1271	.1251	.1230	.1210	.1190	.1170
−1.0	.1587	.1562	.1539	.1515	.1492	.1469	.1446	.1423	.1401	.1379
−.9	.1841	.1814	.1788	.1762	.1736	.1711	.1685	.1660	.1635	.1611
−.8	.2119	.2090	.2061	.2033	.2005	.1977	.1949	.1922	.1894	.1867
−.7	.2420	.2389	.2358	.2327	.2297	.2266	.2236	.2206	.2177	.2148
−.6	.2743	.2709	.2676	.2643	.2611	.2578	.2546	.2514	.2483	.2451
−.5	.3085	.3050	.3015	.2981	.2946	.2912	.2877	.2843	.2810	.2776
−.4	.3446	.3409	.3372	.3336	.3300	.3264	.3228	.3192	.3156	.3121
−.3	.3821	.3783	.3745	.3707	.3669	.3632	.3594	.3557	.3520	.3483
−.2	.4207	.4168	.4129	.4090	.4052	.4013	.3974	.3936	.3897	.3859
−.1	.4602	.4562	.4522	.4483	.4443	.4404	.4364	.4325	.4286	.4247
−.0	.5000	.4960	.4920	.4880	.4840	.4801	.4761	.4721	.4681	.4641
0	.5000	.5040	.5080	.5120	.5160	.5199	.5239	.5279	.5319	.5359
.1	.5398	.5438	.5478	.5517	.5557	.5596	.5636	.5675	.5714	.5753
.2	.5793	.5832	.5871	.5910	.5948	.5987	.6026	.6064	.6103	.6141

Table A.2 Values of the Standard Normal Distribution Function (*continued*)

z	0	1	2	3	4	5	6	7	8	9
.3	.6179	.6217	.6255	.6293	.6331	.6368	.6406	.6443	.6480	.6517
.4	.6554	.6591	.6628	.6664	.6700	.6736	.6772	.6808	.6844	.6879
.5	.6915	.6950	.6985	.7019	.7054	.7088	.7123	.7157	.7190	.7224
.6	.7257	.7291	.7324	.7357	.7389	.7422	.7454	.7486	.7517	.7549
.7	.7580	.7611	.7642	.7673	.7703	.7734	.7764	.7794	.7823	.7852
.8	.7881	.7910	.7939	.7967	.7995	.8023	.8051	.8078	.8106	.8133
.9	.8159	.8186	.8212	.8238	.8264	.8289	.8315	.8340	.8365	.8389
1.0	.8413	.8438	.8461	.8485	.8508	.8531	.8554	.8577	.8599	.8621
1.1	.8643	.8665	.8686	.8708	.8729	.8749	.8770	.8790	.8810	.8830
1.2	.8849	.8869	.8888	.8907	.8925	.8944	.8962	.8980	.8997	.9015
1.3	.9032	.9049	.9066	.9082	.9099	.9115	.9131	.9147	.9162	.9177
1.4	.9192	.9207	.9222	.9236	.9251	.9265	.9278	.9292	.9306	.9319
1.5	.9332	.9345	.9357	.9370	.9382	.9394	.9406	.9418	.9430	.9441
1.6	.9452	.9463	.9474	.9484	.9495	.9505	.9515	.9525	.9535	.9545
1.7	.9554	.9564	.9573	.9582	.9591	.9599	.9608	.9616	.9625	.9633
1.8	.9641	.9648	.9656	.9664	.9671	.9678	.9686	.9693	.9700	.9706
1.9	.9713	.9719	.9726	.9732	.9738	.9744	.9750	.9756	.9762	.9767
2.0	.9772	.9778	.9783	.9788	.9793	.9798	.9803	.9808	.9812	.9817
2.1	.9821	.9826	.9830	.9834	.9838	.9842	.9846	.9850	.9854	.9857
2.2	.9861	.9864	.9868	.9871	.9874	.9878	.9881	.9884	.9887	.9890
2.3	.9893	.9896	.9898	.9901	.9904	.9906	.9909	.9911	.9913	.9916
2.4	.9918	.9920	.9922	.9925	.9927	.9929	.9931	.9932	.9934	.9936
2.5	.9938	.9940	.9941	.9943	.9945	.9946	.9948	.9949	.9951	.9952
2.6	.9953	.9955	.9956	.9957	.9959	.9960	.9961	.9962	.9963	.9964
2.7	.9965	.9966	.9967	.9968	.9969	.9970	.9971	.9972	.9973	.9974
2.8	.9974	.9975	.9976	.9977	.9977	.9978	.9979	.9979	.9980	.9981
2.9	.9981	.9982	.9982	.9983	.9984	.9984	.9985	.9985	.9986	.9986
3.0	.9987	.9990	.9993	.9995	.9997	.9998	.9998	.9999	.9999	.9999

From Lindgren, B.W., *Statistical Theory,* Macmillan, New York, 1993. (With permission.)

Table A.3 The t Distribution

$$P(T \le t) = \int_{-\infty}^{t} \frac{\Gamma[(r + 1)/2]}{\sqrt{\pi r}\,\Gamma(r/2)\,(1 + w^2/r)^{(r+1)/2}}\, dw$$

$$[P(T \le -t) = 1 - P(T \le t)]$$

	P(T ≤ t)				
r	0.90	0.95	0.975	0.99	0.995
1	3.078	6.314	12.706	31.821	63.657
2	1.886	2.920	4.303	6.965	9.925
3	1.638	2.353	3.182	4.541	5.841
4	1.533	2.132	2.776	3.747	4.604
5	1.476	2.015	2.571	3.365	4.032
6	1.440	1.943	2.447	3.143	3.707
7	1.415	1.895	2.365	2.998	3.499
8	1.397	1.860	2.306	2.896	3.355
9	1.383	1.833	2.262	2.821	3.250
10	1.372	1.812	2.228	2.764	3.169
11	1.363	1.796	2.201	2.718	3.106
12	1.356	1.782	2.179	2.681	3.055
13	1.350	1.771	2.160	2.650	3.012
14	1.345	1.761	2.145	2.624	2.977

Table A.3 The t Distribution (*continued*)

			$P(T \leq t)$		
r	0.90	0.95	0.975	0.99	0.995
15	1.341	1.753	2.131	2.602	2.947
16	1.337	1.746	2.120	2.583	2.921
17	1.333	1.740	2.110	2.567	2.898
18	1.330	1.734	2.101	2.552	2.878
19	1.328	1.729	2.093	2.539	2.861
20	1.325	1.725	2.086	2.528	2.845
21	1.323	1.721	2.080	2.518	2.831
22	1.321	1.717	2.074	2.508	2.819
23	1.319	1.714	2.069	2.500	2.807
24	1.318	1.711	2.064	2.492	2.797
25	1.316	1.708	2.060	2.485	2.787
26	1.315	1.706	2.056	2.479	2.779
27	1.314	1.703	2.052	2.473	2.771
28	1.313	1.701	2.048	2.467	2.763
29	1.311	1.699	2.045	2.462	2.756
30	1.310	1.697	2.042	2.457	2.750

This table is taken from Table III of Fisher and Yates: *Statistical Tables for Biological, Agricultural, and Medical Research,* published by Longman Group Ltd., London, 1974 (previously published by Oliver and Boyd, Edinburgh). (With permission).

APPENDIX B: LIST OF UNSAFE ACTS (CHAPTER 4)

Table B.1 List of Unsafe Acts

General code		Unsafe acts
0.		Operating without authority, failure to secure or warn
	00	Starting, stopping, using, operating, moving without authority
	00.1	Cleaning personal clothing in degreaser
	00.2	Using unauthorized solvent for cleaning parts, hands, etc. (Carbon tetrachloride, etc.)
	01	Starting, stopping, using, operating, moving, etc., without giving proper signal
	01.1	Power truck operator failing to signal when passing through blind pedestrian doors
	01.2	Walking through two-way blind pedestrian door without giving indication to oncoming persons
	02	Failing to lock, block, or secure vehicles, switches, valves, press rams, other tools, materials, or equipment against unexpected motion, flow of electric current, steam, etc.
	02.1	Failing to secure material on flats, power trucks
	02.2	Changing cutter on milling machine while machine is running
	02.3	Holding part by hand while drilling (no jig)
	02.4	Failing to secure material in jig
	02.5	Failing to secure jig or fixture on table — no backstop, loose fastenings, etc.
	02.6	Failing to hold part securely while burring, using bench machine
	02.7	Unintentionally tripping punch press
	02.8	Failing to secure drill in chuck
	02.9	Failing to secure vehicle, power truck, etc. against unexpected motion
	03	Failing to shut off equipment not in use
	03.1	Leaving machine unattended while it is running
	05	Failing to place warning signs, signals, tags, etc.
	05.1	Stringing air hose, cords, wire, etc., across aisle without marking or other indications of hazard
	05.2	Failing to label hot parts removed from oven

Table B.1 List of Unsafe Acts (*continued*)

General code	Unsafe acts

1. Operating or working at unsafe speed
   - 10 Running
     - 10.1 Running in or around plant
   - 11 Feeding or supplying too rapidly
     - 11.1 Operating machine at excessive speed
   - 12 Driving too rapidly
     - 12.1 Speeding or improper handling of power truck
     - 12.2 Speeding with automobile in parking lot (collisions)

2. Making safety devices inoperative
   - 20 Removing safety devices
     - 20.1 Removing guard on milling machine
     - 20.2 Grinding without eye shield in place
     - 20.3 Removing guard or making guard inoperative on punch press, welding machine, other machines
     - 20.4 Removing guard (shield) from automatic screw machine
     - 20.5 Removing guard from chuck motor
     - 20.6 Removing guard from circular saw
   - 21 Blocking, plugging, tying, etc. of safety devices
     - 21.1 Reaching behind guard on punch press, welding machine ("cheating" the guard — circumventing the guard)
     - 21.2 Circumventing the glass shield on grinding wheel
   - 23 Misadjusting safety device
     - 23.1 Guard on milling machine not properly adjusted
     - 23.2 Work rest improperly adjusted on grinding wheel
     - 23.3 Guard on punch press, welding machine, or other machine improperly adjusted or inadequate (opening too wide, guard loose, etc.)

3. Using unsafe equipment, hands instead of equipment, or equipment unsafely
   - 30 Using defective equipment
     - 30.1 Using makeshift fastener (wire) to fasten guard or shield in place
     - 30.2 Using improperly maintained tools and equipment
     - 30.3 Using weak or damaged flats
     - 30.4 Using worn nuts, bolts, threads, wrenches, etc., on jigs, fixtures, vises, machines, etc.
     - 30.5 Working with exposed, unguarded work light
     - 30.6 Using makeshift, unstable stand for holding stock pans
   - 31 Unsafe use of equipment
     - 31.1 Using air hose to blow chips off machine, out of jig, etc.
     - 31.2 Using air hose to clean body, clothes, hair, etc.
     - 31.3 Wiring air hose in "on" position — fastening air hose down on drill press in open position
     - 31.4 Unsafe improvising — using improper or makeshift tools or equipment
     - 31.5 Cleaning under dies, removing parts from under dies, putting hands under punch press dies while machine is running
     - 31.6 Excessively high *psi* pressure on air hose
     - 31.7 Grinding material too heavy for size of grinding wheel
   - 32 Using hands instead of hand tools, etc.
     - 32.1 Handling metal chips by hand
     - 32.2 Removing frozen drill from jig bushing by hand
     - 32.3 Holding part by hand while drilling, tapping (no jig)

Table B.1 List of Unsafe Acts (*continued*)

General code	Unsafe acts
33	Gripping objects insecurely, taking wrong hold of objects
33.1	Hand slipping from part, striking belt sander, drill, etc.
33.2	Dropping material, tools, pans, etc. onto floor, body
33.3	Grabbing spindle of drill press by hand while it is rotating — attempting to stop spindle by hand with power on
33.4	Burring part — hand or tool slipped, striking drill, tool

4. Unsafe loading, placing, mixing, combining, etc.
- 40 Overloading
  - 40.1 Overloading flats, power trucks
  - 40.2 Overloading stock pans
- 41 Crowding
  - 41.1 Machines too close together — inadequate operator work space
- 42 Lifting or carrying too heavy loads
- 43 Arranging or placing objects or materials unsafely
  - 43.1 Placing scrap material on floor around machine
  - 43.2 Unstable stacking of stock pans on makeshift stands
  - 43.3 Storing materials in aisles, workerstrip, bumping into material while walking
  - 43.4 Unstable stacking of stock pans on floor
  - 43.5 Storing material, equipment in designated aisles — struck by power trucks or other vehicles
  - 43.6 Keeping tools on floor
  - 43.7 Placing stock rack for automatic screw machine in designated aisle
  - 43.8 Stringing air hose, cords, wire, etc. across aisle
  - 43.9 Holding part by hand while drilling, tapping — no jig
  - 43.10 Holding jig by hand while drilling, tapping — no backstop
  - 43.11 Throwing plastic spoons on floor in cafeteria and in work areas — slips, falls
  - 43.12 Chips retained in jig misaligns jig resulting in broken drill on drill press
  - 43.13 Misaligning material in punch press
- 45 Introducing objects or materials unsafely
  - 45.1 Spilling oil on floor

5. Taking unsafe position or posture
- 54 Lifting with bent back, while in awkward position, etc.
  - 54.1 Improper lifting procedure
- 56 Exposure on vehicle right-of-way
  - 56.1 Operating power truck on route used by pedestrians — truck strikes pedestrians (entering aisle from restrooms, etc.)
- 59 Miscellaneous unsafe positions or postures
  - 59.1 Touching hot light fixture
  - 59.2 Exposing parts of body to flying chips from milling, drilling, etc.
  - 59.3 Splash burn from welding — touching hot welding rod
  - 59.4 Assuming awkward position while tightening nut
  - 59.5 Reaching too far, losing balance, fall from chair
  - 59.6 Pinching finger in drill jig while closing jig
  - 59.7 Mashing finger on hands while moving flats, pans, etc. (not dropped)
  - 59.8 Stumbling over wooden platform supporting machinery

Table B.1 List of Unsafe Acts (*continued*)

General code		Unsafe acts
6.		Working on moving or dangerous equipment
	61	Cleaning, oiling, adjusting, etc. of moving equipment
	61.1	Changing cutter on milling machine while machine is running
	61.2	Adjusting machine while it is in motion (gauging work, removing part, etc.)
	61.3	Cleaning chips from milling, drilling, and other machines while machines are in motion
	61.4	Cleaning under dies, removing parts, putting hands under dies of punch press with power on
	61.5	Reaching inside of automatic screw machine to remove part while machine is in motion
	61.6	Putting drill in drill press chuck while spindle is turning (standard chuck)
	63	Working on electrically charged equipment
	63.1	Electric shock from machine, faulty wiring, other causes (power-on maintenance, etc.)
7.		Distracting, teasing, abusing, startling, etc.
	70	Calling, talking, or making unnecessary noise
	70.1	Distraction from fellow worker on supervisor
	70.2	Horseplay — making loud, startling noise behind operator (drop pans, etc.)
	71	Throwing material
	71.1	Horseplay — throwing small pieces of stock at other operators
	72	Teasing, abusing, startling, etc.
	72.1	Tickling operator in ribs or under arms
	72.2	Pinching female operator
	72.3	Intentionally bumping operator using drill press, grinding wheel, etc.
	73	Practical joking, etc.
	73.1	Opening interlocked door on gear box of milling machine while operator is away — machine fails to start when operator returns
	73.2	Miscellaneous horseplay: (1) hooking can of oil behind worker, (2) putting chips on seat of chair, (3) making loud noises, (4) loosening drill, (5) turning off machine while in use, (6) hiding jig, piece parts, etc.
8.		Failure to use safe attire or personal protective devices
	80	Failing to wear goggles, gloves, masks, aprons, shoes, leggings, etc.
	80.1	Operating milling machine without safety glasses
	80.2	Operating a punch press without safety glasses
	80.3	Operating a screw machine or engine lathe without safety glasses
	80.4	Operating a drill press without safety glasses
	80.5	Operating a grinding machine, welding machine, other machine without safety glasses
	80.6	Outsiders walking through designated eye hazard area without safety glasses
	80.7	Oil or chemical splash in eye (no goggles or goggles unsuited for splash protection)
	80.8	Wearing defective or unsafe goggles or goggles unsuited for impact protection (glasses without side shields)
	80.9	Handling burred or sharp-edged stock without gloves
	80.10	Handling hot machined parts with unprotected hands

Table B.1 List of Unsafe Acts (*continued*)

General code		Unsafe acts
	80.11	Handling hot parts from oven without gloves
	80.12	Gloves worn while handling sheet steel or other materials that are unsuited for protection from sharp edges of material
	80.13	Failing to wear hard-toe safety shoes while handling heavy material
	80.14	Failing to wear ear plugs in noise hazard areas
81		Wearing high heels, loose hair, long sleeves, loose clothing, etc.
	81.1	Loose, long hair around revolving machinery
	81.2	Wearing loose clothes, long sleeves, rings, etc. around rotating machinery
	81.3	Wearing gloves while working with revolving machines (drills, milling machines, lathes, etc.)
	81.4	Wearing gloves while grinding, buffing, etc.
	81.5	Wearing high heels, tripping, etc., in areas containing holes, depressions, protrusions, etc. in floor
99		Not elsewhere classified — jig heats up during drilling operation, burning operator (no coolant)

From Tarrents, W. E., The Measurement of Safety Performance, Garland STPM Press, New York, 1980. (With permission.)

APPENDIX C: SAFETY INSPECTION CHECKLIST
(CHAPTER 5)

The result of safety inspections can be analyzed with a view to identify areas that are more hazardous than others. It is a common practice to color code a map of the plant as high hazard, moderate hazard, and low hazard areas. Safety inspections are conducted with varying frequencies of visits to areas with different degrees of hazard. The most frequent inspections should be carried out in high hazard areas. The next priority should be given to moderate hazard areas. Low hazard areas should be inspected less frequently than the other two areas.

C.1   GENERAL INDUSTRY SAFETY AND HEALTH CHECKLIST

This checklist is based on the General Industrial Safety and Health List. It proceeds a comprehensive listing of factors that must be considered while conducting an inspection. A scoring scheme for the inspection is suggested.

Failure to properly define the unsafe condition is perhaps the single cause of trouble in improving plant safety. In this checklist, every question pertains to a work center. Score 1 if this work center exists and 0 otherwise (this is X_i); under each work center with a score of

1, evaluate each factor and give a score range between 0 to 5, 0 for poor and 5 for excellent (this is S_{ij}). The weight of each question may be changed to suit the plant type. Total score of work center i is S_i and n represents the number of question under work center i.

Once the importance of various factors has been determined, fill in the location boxes to determine how well they fit with what you consider important. Thus, time spent in evaluating unimportant/unpertinent factors may be minimized.

No checklist guarantees an improvement in the plant safety by itself. Such lists can only aid the safety inspection teams in conducting appraisals.

			Not
1.	**Abrasive Blasting —**		
	Work Center (i)	Applicable	applicable
	(X_i)	☐	☐

a. Are blast-cleaning nozzles equipped with an operating valve that must be held open manually?

$S_{i,j}$ _____

Is a support provided for each nozzle on which the nozzle may be mounted when not in use?

$S_{i,j}$ _____

b. Is the concentration of respirable dust or fumes in the breathing zone of the abrasive-blasting operator or any other worker below the specified levels?

$S_{i,j}$ _____

c. Are blast-cleaning enclosures exhaust ventilated in such a way that a continuous inward flow of air is maintained at all openings in the enclosure during the blasting operation?

$S_{i,j}$ _____

d. Is the air for abrasive-blasting respirators free of harmful quantities of contaminants?

$S_{i,j}$ _____

S_i _____

2. Abrasive grinding (X_i) ☐ ☐

a. Are abrasive wheels used only on machines provided with safety guards? The following exceptions are allowed:
- Wheels used for internal work while within the work being ground
- Mounted wheels, used in portable operations, 2 in. and smaller in diameter

$S_{i,j}$ _____

b. Are abrasive wheel bench and stand grinders provided with safety guards which cover the spindle ends, nut, and flange? The following exceptions are allowed:
- Safety guards on all operations where the work provides a suitable measure of protection to the operator may be constructed that the spindle end, nut, and outer flange are exposed.
- Where the nature of the work is such as to entirely cover the side of the wheel, the side covers of the guard may be omitted.
- The spindle end, nut, and outer flange may be exposed on machine designed as portable saws.

$S_{i,j}$ _____

c. Is an adjustable work-rest of rigid construction used to support the work on fixed base, offhand grinding machines? Are work rests kept adjusted closely to the wheel with a maximum opening of $\frac{1}{8}$ inch? Is the work-rest securely clamped after each adjustment?

$S_{i,j}$ _____

d. While performing dry grinding, are one suitable hood or enclosures provided that are connected to exhaust systems to control airborne contaminants?

$S_{i,j}$ _____

e. Are machines designed for a fixed location securely anchored to prevent walking or moving?

$S_{i,j}$ _____

S_i _____

3. Air Receivers, (X_i) □   □
Compressed

a. Are air receivers supported with sufficient clearance to permit a complete external inspection and to avoid corrosion of external surfaces?

$S_{i,j}$ _____

b. Are air receivers installed so that drains, handholes, and manholes are easily accessible?

$S_{i,j}$ _____

c. Is every air receiver equipped with an indicating pressure gauge so located as to be readily visible and with one or more spring-loaded safety valve?

$S_{i,j}$ _____

S_i _____

4. Air Tools (X_i) ☐ ☐

a. For portable tools, is a tool retainer installed on each piece of utilization equipment, which without such a retainer, may eject the tool?

$S_{i,j}$ _____

b. Are hose and hose connections used for conducting compressed air to utilization equipment designed for the pressure and service to which they are subject?

$S_{i,j}$ _____

S_i _____

5. Aisles and Passageways (X_i) ☐ ☐

a. Where mechanical handling equipment is used, is sufficient safe clearance allowed for aisles at loading docks, through doorways, and whenever turns or passage are made?

$S_{i,j}$ _____

b. Are aisles and passageways kept clear and in good repair with no obstructions across or in aisles that could create hazards?

$S_{i,j}$ _____

c.   Are permanent aisles and pas-
       sageways appropriately
       marked?

$S_{i,j}$ _____

S_i _____

6.   Belt Sanding Machines (X_i)   ☐   ☐
       (Woodworking)

a.   Are belt-sanding machines pro-
       vided with guards at each nip
       point where the sanding belt
       runs onto a pulley?

$S_{i,j}$ _____

b.   Is the unused run of the sanding
       belt guarded against accidental
       contact?

$S_{i,j}$ _____

S_i _____

7.   Boiler (X_i)   ☐   ☐

a.   Is boiler inspection and ap-
       proval, on an annual basis, by
       a recognized boiler inspection
       service satisfactory evidence of
       acceptable installation and
       maintenance?

$S_{i,j}$ _____

b.   Is valid boiler inspection certifi-
       cate, bearing the signature of
       the authorized inspector and the
       date of the last inspection, con-
       spicuously posted?

$S_{i,j}$ _____

c.   Are boilers equipped with an ap-
       proved means of determining

the water level such as water column, gauge glass, or try cocks? Are gauge glasses and water columns guarded to prevent breakage?

$$S_{i,j}$$ _____

$$S_i$$ _____

8. Calendars, Mills, and Rolls (X_i) ☐ ☐

a. Is a safety trip-type bar, rod, or cable to activate an emergency stop-switch installed on calendars, rolls, or mills to prevent persons or parts of the body from being caught between the rolls?

$$S_{i,j}$$ _____

b. Is a fixed guard across the front and one across the back of the mill, approximately 40 in. vertically above the working level and 20 in. horizontally from the crown face of the roll, used where applicable?

$$S_{i,j}$$ _____

$$S_i$$ _____

9. Chains, Cables, Ropes, and so on (Overhead and Gantry Cranes) (X_i) ☐ ☐

a. Are chains, cables, ropes, slings, etc. inspected daily and defective gear removed and repaired or replaced?

$$S_{i,j}$$ _____

b.   Are hoist chains and hoist ropes
       free from kinks or twists and
       wrapped around the load?

$$S_{i,j}$$ _____

c.   Are all U-bolt wire rope clips or
       hoist ropes installed so that the
       U-bolt is in contact with the
       dead end (short or nonload car-
       rying end) of the rope? Are
       clips installed in accordance
       with the clip manufacturer's
       recommendation? Are nuts or
       newly installed clips retight-
       ened after 1 h of use?

$$S_{i,j}$$ _____
$$S_i$$ _____

10.   Chip Guards (X_i) ☐   ☐

a.   Are protective shields and bar-
       riers provided, in operations in-
       volving cleaning with com-
       pressed air, to protect
       personnel against flying chips
       or other such hazards?

$$S_{i,j}$$ _____
$$S_i$$ _____

11.   Chlorinated hydro- (X_i) ☐   ☐
       carbons

a.   Are carbon tetrachloride or other
       chlorinated (halogenated) hy-
       drocarbons used where the air-
       borne concentration exceeds the
       threshold limit value (TLV)
       listed?

$$S_{i,j}$$ _____

b. Are degreasing or other cleaning operations involving chlorinated hydrocarbons so located that vapors from these operations reach or draw into the atmosphere surrounding any welding operations?

$S_{i,j}$ _____

S_i _____

12. Compressed Air (X_i) ☐ ☐

a. Does compressed air used for cleaning purposes exceed 30 psi? Are effective chip guards and personal protective equipment provided when it increases above 30 psi?

$S_{i,j}$ _____

S_i _____

13. Cone Pulleys (Mechanical Power-Transmission Equipment) (X_i) ☐ ☐

a. Are the cone belt and pulleys equipped with a belt shifter so constructed as to adequately guard the nip point of the belt and pulley? If the frame of the belt shifter does not adequately guard the nip point of the belt and the pulley, is the nip point further protected by means of a vertical guard placed in front of the pulley and extending at least to the top of the largest step of the cone?

$S_{i,j}$ _____

S_i _____

14.   Conveyors (X_i) ☐   ☐

a.   Are conveyors installed within 7
       ft of the floor or walkways
       provided with crossovers at
       aisles of other passageways?
                                      $S_{i,j}$ _____

b.   Where conveyors 7 ft or more
       above the floor pass over
       working areas, aisles, or tho-
       roughfares, are suitable guards
       provided to protect personnel
       from the hazard of falling ma-
       terials?
                                      $S_{i,j}$ _____

c.   Are open hoppers and chutes
       guarded by standard railings
       and toeboards or by some other
       comparable safety device?
                                      $S_{i,j}$ _____
                                      S_i _____

15.   Cranes and Hoists (X_i) ☐   ☐
       **(Overhead and
       Gantry)**

a.   Are all functional operating
       mechanisms, air and hydraulic
       systems, chains, rope slings,
       hooks, and other lifting equip-
       ment inspected daily?
                                      $S_{i,j}$ _____

b.   Is complete inspection of the
       crane performed at intervals de-
       pending on its activity, severity
       of service, and environment?
                                      $S_{i,j}$ _____

c. Does the overhead crane have stops at the limit of travel of the trolley, bridge, and trolley bumpers or equivalent automatic services, and rail sweeps on the bridge trucks?

$S_{i,j}$ _____

d. Is the rated load of the crane plainly marked on each side of the crane, and if the crane has more than one hoisting unit, does each hoist have its rated load marked on it or its load block, and is this marking clearly legible from the ground or floor?

$S_{i,j}$ _____

S_i _____

16. Cylinders and Compressed Gas (Used in Welding) (X_i) ☐ ☐

a. Are compressed gas cylinders kept away from excessive heat, stored where they will not be damaged or knocked over by passing or falling objects, and stored at least 20 ft away from highly combustible materials?

$S_{i,j}$ _____

b. Where a cylinder is designed to accept a valve protection cap, are caps in place except when the cylinder is in use or is connected for use?

$S_{i,j}$ _____

c.   Are acetylene cylinders stored in
        a vertical, valve-end-up posi-
        tion only?

                        $S_{i,j}$ _____

d.   Are oxygen cylinders in storage
        separated from fuel-gas cylin-
        ders or combustible materials
        (especially oil or grease) a
        minimum distance of 20 ft or
        by a noncombustible barrier at
        least 5 ft high, having a fire-
        resistance rating of at least 1 h?

                        $S_{i,j}$ _____
                        S_i _____

**17.   Dip Tanks Containing (X_i) ☐   ☐
        Flammable or Com-
        bustible Liquid**

a.   Are dip tanks of over 150 gal
        capacity or 10 ft² in liquid sur-
        face area equipped with a prop-
        erly trapped overflow pipe
        leading to a safe location out-
        side the building?

                        $S_{i,j}$ _____

b.   Are open flames, spark-produc-
        ing devices, or heated surfaces
        having a temperature sufficient
        to ignite vapors in or within 20
        ft of any vapor area? Is electri-
        cal wiring and equipment in
        any vapor area of the explo-
        sion-proof type? Is there elec-
        trical equipment in the vicinity
        of dip tanks, associated drain
        boards, or drying operations
        that are subject to splashing or
        dripping?

                        $S_{i,j}$ _____

c.   Are all dip tanks (except hardening and tempering tanks) exceeding 150-gal liquid capacity,
having a liquid surface area exceeding 4 ft^2, protected with at
least one of the following automatic extinguishing facilities:
water spray system, foam system, carbon dioxide system,
dry chemical system, or automatic dip tank cover? This pro-
vision applies to hardening and
tempering tanks having a liquid
surface area of 25 ft^2 or more
of a capacity of 500 gal or
more.

$S_{i,j}$ _____

S_i _____

18.   Dockboards (X_i) ☐   ☐

a.   Are dockboards strong enough
to carry the load imposed on
them?

$S_{i,j}$ _____

b.   Are portable dockboards anchored or equipped with devices that will prevent their
slipping? Do they have hand
holds or other effective means
to allow safe handling?

$S_{i,j}$ _____

c.   Are positive means provided to
prevent railroad cars from
being moved while dockboards
are in position?

$S_{i,j}$ _____

S_i _____

19. **Drains for Flammable and Combustible Liquids** (X_i) ☐ ☐

a. Are emergency drainage systems provided to direct flammable liquid leakage and fire protection water to a safe location?

$S_{i,j}$ _____

b. Are emergency drainage systems for flammable liquids that are connected to public sewers or discharged into public waterways equipped with traps or separators?

$S_{i,j}$ _____

S_i _____

20. **Drill Presses** (X_i) ☐ ☐

a. Is the V-belt drive of all drill presses, including the usual front and rear pulleys, guarded to protect the operator from contact or breakage?

$S_{i,j}$ _____

S_i _____

21. **Electrical Installations** (X_i) ☐ ☐

a. Electrical installations and all new utilization equipment installed before March 15, 1972, shall be installed in accordance with the current regulations. For locations in the U.S., refer to National Electrical Code, NFPA 70-1971; ANSI CI-1971 (rev. of 1968).

$S_{i,j}$ _____

S_i _____

22. Emergency Flushing, (X_i) ☐ ☐
Eyes, and Body

a. Where the eyes or body of any
person is exposed to injurious
corrosive materials, are suitable
facilities for quick drenching or
flushing of the eyes and body
provided within the work area
for immediate emergency use?

$S_{i,j}$ _____

S_i _____

23. Exits (X_i) ☐ ☐

a. Are buildings designed for human occupancy provided with
exits sufficient to permit the
prompt escape of occupants in
case of emergency?

$S_{i,j}$ _____

b. Where occupants may be endangered by the blocking of any
single egress due to fire or
smoke, are there at least two
means of egress remote from
each other?

$S_{i,j}$ _____

c. Are exits and the ways of approach and travel from exits,
well-maintained so that they
are unobstructed and are accessible at all times?

$S_{i,j}$ _____

d. Do all exits open directly to the
street or other open space that
gives safe access to a public
way?

$S_{i,j}$ _____

e. Do exit doors serving more than 50 people, or at high hazard areas, swing in the direction of travel?

$S_{i,j}$ _____

f. Are exits marked by readily visible, illuminated exit signs? Are exit signs distinctive in color and provide contrast with surroundings? Is the word "EXIT" of plainly legible letters, not less than 6 in. high?

$S_{i,j}$ _____
S_i _____

24. Explosives and Blasting Agents (X_i) ☐ ☐

a. Are all explosives kept in approved magazines?

$S_{i,j}$ _____

b. Are stored packages of explosives laid flat with top side up. Is black powder, when stored in magazines with other explosives, stored separately?

$S_{i,j}$ _____

c. Are smoking, matches, open flames, spark-producing devices, and firearms (except firearms carried by guards) permitted inside of or within 50 ft of magazines? Is the land surrounding a magazine kept clear of all combustible materials for a distance of at least 25 ft? Are combustible materials stored within 50 ft of magazines?

$S_{i,j}$ _____
S_i _____

25. **Fan Blades** (X_i) ☐ ☐

a. When the periphery of the blades of a fan is less than 7 ft above the floor or working level, are the blades guarded? Does the guard have openings no larger than $1/2$ in. (The use of concentric rings with space between them, not exceeding $1/2$ in., is acceptable, provided they are adequately supported).

$$S_{i,j} \underline{\hspace{5cm}}$$
$$S_i \underline{\hspace{5cm}}$$

26. **Fire Protection** (X_i) ☐ ☐

a. Are portable fire extinguishers suitable to the conditions and hazards involved provided and maintained in an effective operating condition?

$$S_{i,j} \underline{\hspace{5cm}}$$

b. Are portable fire extinguishers conspicuously located and mounted where they will be readily accessible? Are extinguishers obstructed from view?

$$S_{i,j} \underline{\hspace{5cm}}$$

c. Are portable fire extinguishers given maintenance service at least once a year with a durable tag security attached to show the maintenance or recharge date?

$$S_{i,j} \underline{\hspace{5cm}}$$

d.   In storage areas, clearance between sprinkler system defec-
tors and top of storage varies
with the type of storage. For
combustible material stored
over 15 ft, but not more than
21 ft, high in solid piles, or
over 12 ft, but not more than
21 ft, high in piles that contain
horizontal channels, is the minimum clearance 36 in.? Is the
minimum clearance for smaller
piles or for noncombustible
materials 18 in.?

$S_{i,j}$ _____

S_i _____

27.   **Flammable Liquids In-** (X_i) ☐   ☐
cidental to Principal
Business

a.   Are flammable liquids kept in
covered containers when not
actually in use?

$S_{i,j}$ _____

b.   Does the quantity of flammable
or combustible liquid that is located outside of an inside storage room or storage cabinet in
any one fire area of a building
exceed:
• 25 gal of class 1A liquids in
containers
• 120 gal of class 1B, 1C, II,
or III liquids in containers
• 660 gal of class 1B, 1C, II,
or III in a single portable
tank

$S_{i,j}$ _____

c.   Are flammable and combustible liquids drawn from or transferred into containers within a building only through a closed piping system, from safety cans, by means of a device drawing through the top, or by gravity through an approved self-closing valve? Is transferring by means of air pressure prohibited?

$S_{i,j}$ _____

d.   Do inside storage rooms for flammable and combustible liquids of fire-resistive construction, have self-closing fire doors at all openings, 4-in. sills or depressed floors, a ventilation system that provides at least 6 air changes within the room per hour, and, in areas used for storage of class I liquids, electrical wiring approved for use in hazardous locations?

$S_{i,j}$ _____

e.   Are outside storage areas graded in such a manner to divert spills away from buildings or other exposures, or surrounded by curbs or dikes at least 6 in. high with appropriate drainage to a safe location for accumulated liquids? Is the area protected against tampering or trespassing, where necessary, and kept free of weeds debris, and other combustible material not necessary to the storage?

$S_{i,j}$ _____

f.   Are areas where flammable liq-
       uids with flashpoints below
       100°F are used ventilated at a
       rate of not less than 1 ft³/min/
       ft² of solid floor area?

                              $S_{i,j}$ _____
                              S_i _____

28.   **Floors, General Condi-** (X_i) ☐   ☐
       tions

a.   Are all floor surfaces kept clean,
       dry, and free from protruding
       nails, splinters, loose boards,
       holes, or projections?

                              $S_{i,j}$ _____

b.   Where wet processes, is drain-
       age well-maintained, and false
       floors, platforms, mats, or
       other dry standing places pro-
       vided?

                              $S_{i,j}$ _____
                              S_i _____

29.   **Floor Loading Limit** (X_i) ☐   ☐

a.   In buildings used for mercantile,
       business, industrial, or storage
       purposes, are all floors posted
       to show maximum safe floor
       loads?

                              $S_{i,j}$ _____
                              S_i _____

30.   **Floor Openings, Hatch-** (X_i) ☐   ☐
       ways, Open Sides,
       Etc.

a.   Are floor openings requiring ac-
       cess by personnel, such as

stairway openings and ladderway openings, guarded by a
standard railing on all exposed
sides except at the access
point? Is access to ladderway
openings further guarded so
that a person cannot walk directly into the opening? Are
other floor openings guarded
by a suitable covering, and further guarded by a removable
standard railing, when the covering is removed or constantly
attended by someone? Are skylight openings guarded by a
standard skylight screen or
fixed standard?

$S_{i,j}$ _____

b. Are open-sided floors, platforms, etc. 4 ft or more above
the adjacent floor or ground
level guarded by a standard
railing on all open sides, except where there is an entrance
to a ramp, stairway, or fixed
ladder?

$S_{i,j}$ _____
S_i _____

31. Foot Protection (X_l) ☐ ☐

a. Does safety-toe footwear meet
the requirements of the pertinent standards?

$S_{i,j}$ _____
S_i _____

32.   Fork-Lift Trucks (X_i) ☐   ☐

a.   Do all new fork-lift trucks bear
       a label indicating approval in
       accordance with the current
       regulations?

                              $S_{i,j}$ _____

b.   Are high-lift rider trucks
       equipped with a substantial
       overhead guard unless operat-
       ing conditions do not permit?

                              $S_{i,j}$ _____

c.   Are fork-lift trucks equipped
       with a vertical load backrest
       extension when the type of
       load presents a hazard to the
       operator?

                              $S_{i,j}$ _____

d.   Are the brakes of highway
       trucks set and wheel chocks
       placed under the rear wheels to
       prevent the truck from rolling
       while they are boarded with
       fork-lift trucks?

                              $S_{i,j}$ _____

e.   Are wheel stops or other recog-
       nized protection provided to
       prevent railroad cars from mov-
       ing while they are boarded with
       fork-lift trucks?

                              $S_{i,j}$ _____
                              S_i _____

33.   General Duty Clause (X_i) ☐   ☐

a.   Hazardous conditions or prac-
       tices not covered in an OSHA

standard may be covered under Section 5(a)(1) of OSHA which states: "Each employer shall furnish to each employee the feeling that their place of employment is free from recognized hazards that is causing or is likely to cause death or serious physical harm to him".

$S_{i,j}$ _____

S_i _____

34. Guards and their Construction (X_i) ☐ ☐

a. Are guards for mechanical power transmission equipment made of metal? (Wood guards may be used in the woodworking and chemical industries, in industries where atmospheric conditions would rapidly deteriorate metal guards, or where temperature extremes make metal guards undesirable.)

$S_{i,j}$ _____

S_i _____

35. Head Protection (X_i) ☐ ☐

a. Does head-protective equipment meet the requirements of ANSI Z89.1-1967, Requirements for Industrial Head Protection or its equivalent standard?

$S_{i,j}$ _____

S_i _____

36. Hand Tools (X_i) ☐ ☐

a. Are the hand tools and equipment used by employees in

safe condition (including tools and equipment that may be furnished by employees)?

$S_{i,j}$ _____

S_i _____

37. Hooks, Cranes, and Hoists (see Cranes, and Hoists, No. 15)

38. Housekeeping (X_i) ☐ ☐

a. Are all places of employment, passageways, storerooms, and service rooms kept clean and orderly and in a sanitary condition?

$S_{i,j}$ _____

S_i _____

39. Jointers (Woodworking) (X_i) ☐ ☐

a. Is each hand-fed planer and jointer with a horizontal head equipped with a cylindrical cutting head? Is the opening in the table kept as small as possible?

$S_{i,j}$ _____

b. Does each hand-fed jointer with a horizontal cutting head have an automatic guard that covers the section of the head on the working side of the fence or gauge?

$S_{i,j}$ _____

c. Does a jointer guard automatically adjust itself to cover the unused portion of the head and

remain in contact with the material at all times?

$S_{i,j}$ _____

d. Does each hand-fed jointer horizontal cutting head have a
guard that covers the section of
the head back of the gauge or
fence?

$S_{i,j}$ _____
S_i _____

40. Ladders, Fixed (X_i) ☐   ☐

a. Are all fixed ladders designed
for a minimum concentrated
live load of 200 lbs?

$S_{i,j}$ _____

b. Do all rungs have a minimum
diameter of $^3/_4$ in., if metal? If
the ladder is constructed of
metal rungs embedded in concrete and exposed to a corrosive atmosphere, do the rungs
have a minimum diameter of
1 in.? Do wooden ladders have
rungs with a minimum diameter of 1 in.? Are all rungs
spaced uniformly, not more
than 12 in. apart and have a
minimum clear length of 16
in.?

$S_{i,j}$ _____

c. Are metal ladders painted or
treated to resist corrosion or
rusting when the location de-
mands?

$S_{i,j}$ _____

d.   Are cages, wells, or ladder
        safety devices for ladders af-
        fixed to towers, water tanks, or
        chimneys provided on all lad-
        ders more than 20 ft long? Are
        landing platforms provided
        each 30 ft in length, except
        where no cage is provided? Are
        landing platforms provided for
        every 20 ft of length?

$S_{i,j}$ _____

e.   Do tops of cages on fixed lad-
        ders extend 42 ft above the top
        of the landing, unless other ac-
        ceptable protection is provided,
        and the bottom of the cage not
        less than 7 ft, nor more than 8
        ft, above the base of the lad-
        der?

$S_{i,j}$ _____

f.   Do side rails of through or side-
        step ladder extensions extend 3
        ft above parapets and landings?
        For through ladder extensions,
        are the rungs omitted from the
        extension and have no less than
        18 nor more than 24 in. clear-
        ance between rails? (For side-
        step or offset fixed ladder sec-
        tions at landings, the side rails
        and rungs shall be carried to
        the next regular rung beyond or
        above the 3 ft minimum).

$S_{i,j}$ _____
S_i _____

41.   Ladders, Portable   (X_i)   ☐   ☐

a.   Is the maximum length for port-
        able wooden ladders and step-

ladders 20 ft; single straight
ladders 30 ft; 2-section extension ladders 60 ft; sectional
ladders 60 ft; trestle ladders 20
ft; platform stepladders 20 ft;
painter's stepladders, 12 ft; and
masons' ladders, 40 ft?

$S_{i,j}$ _____

b. Is the maximum length for portable metal ladders and single,
straight ladders 30 ft; 2-section
extension ladders 48 ft; over 2-
section extension ladders 60 ft;
stepladders, 20 ft; trestle ladders, 20 ft; and platform stepladders, 20 ft?

$S_{i,j}$ _____

c. Are stepladders equipped with a
metal spreader or locking device of sufficient size and
strength to securely hold the
front and back sections in open
position?

$S_{i,j}$ _____

d. Are ladders maintained in good
condition and defective ladders
withdrawn from service?

$S_{i,j}$ _____

e. Are non-self supporting ladders
erected on a sound base at a 4
to 1 pitch and placed to prevent
slipping?

$S_{i,j}$ _____

f. Does the top of a ladder used to
gain access to a roof extend at

least 3 ft above the point of
contact?

$$S_{i,j} \quad \underline{\hspace{5cm}}$$

g.   Are wooden ladders kept coated
with a suitable protective material?

$$S_{i,j} \quad \underline{\hspace{5cm}}$$

h.   In general industrial use, when
portable metal ladders are used
in areas containing electrical
circuits, are proper safety measures taken?

$$S_{i,j} \quad \underline{\hspace{5cm}}$$
$$S_{i} \quad \underline{\hspace{5cm}}$$

42.   Lighting (X_i) ☐   ☐

a.   Is adequate illumination, depending on the seeing tasks involved, provided and distributed to all areas in accordance
with ANSI Standard A11.1?
(Requirements vary widely, but
a good rule of thumb is 20 to
30 ft candles for services and
50 to 100 ft candles for tasks.)

$$S_{i,j} \quad \underline{\hspace{5cm}}$$
$$S_{i} \quad \underline{\hspace{5cm}}$$

43.   Lunchrooms (X_i) ☐   ☐

a.   Do employees consume food or
beverages in toilet rooms or in
any area exposed to a toxic
material?

$$S_{i,j} \quad \underline{\hspace{5cm}}$$

b.   Are covered receptacles corro-
sion-resistant to disposable ma-

terial provided in lunch areas for disposal of waste food? (The cover may be omitted where sanitary conditions can be maintained without the use of a cover.)

$S_{i,j}$ _____

S_i _____

44. Machine Guarding (X_i) ☐ ☐

a. Are one or more methods of machine guarding provided to protect the operator and other employees in the machine area from hazards such as those created by point of operation, ingoing nip points, rotating parts, and flying chips or sparks?

$S_{i,j}$ _____

S_i _____

45. Machine, Fixed (X_i) ☐ ☐

a. Are machines designed for a fixed location securely anchored to prevent walking or moving?

$S_{i,j}$ _____

S_i _____

46. Mats, Insulating (X_i) ☐ ☐

a. Are motors or controllers operating at more than 150 V to ground, grounded against accidental contact only by location, and, where adjustment or other attendance may be necessary during operations, are suitable

insulating mats or platforms
provided?

$S_{i,j}$ _____

S_i _____

**47.   Medical Services and (X_i) ☐   ☐
First Aid**

a.   Are medical personnel readily
available for advice and consul-
tation on matters of plant
health?

$S_{i,j}$ _____

b.   Are persons trained to render
first aid when a medical facil-
ity for treatment of injured employees is not available in near
proximity to the workplace?

$S_{i,j}$ _____

c.   Are necessary first-aid supplies
readily available?

$S_{i,j}$ _____

S_i _____

48.   Noise Exposure (X_i) ☐   ☐

a.   Is protection against the effects
of occupational noise exposure
provided when the sound levels
exceed those shown in Safety
and Health Standards? Are feasible engineering and/or administrative controls utilized to
keep exposure below the allowable limit?

$S_{i,j}$ _____

b.   When engineering or administrative controls fail to reduce the

noise level to within the allowable levels, is personal protective equipment provided and
used to reduce the noise to an
acceptable level?

$S_{i,j}$ _____

c. Does exposure to impulsive or
impact noise exceed 140-dB
peak sound-pressure level?

$S_{i,j}$ _____

d. In all cases where the sound
levels exceed the values shown
in e, is an effective hearing
conservation program adminis-
tered?

$S_{i,j}$ _____

e. Suggested Permissible Noise Exposure:

Duration per day (h)	Sound level (dBA) slow response
8	90
6	92
4	95
3	97
2	100
1	102
1	105
$^{1}/_{2}$	110
$^{1}/_{4}$ or less	115

$S_{i,j}$ _____
S_i _____

49. Personal Protective (X_i) ☐   ☐
Equipment

a Where there is hazard from processes of environment that may
cause injury or illness to the
employee, is proper personal
protective equipment, including

shields and barriers, provided,
used, and maintained in a sanitary and reliable condition?

$S_{i,j}$ _____

b. Where employees furnish their
own personal protective equipment, does the employer assure
its adequacy and that the equipment is properly maintained
and in a sanitary condition?

$S_{i,j}$ _____
S_i _____

**50.   Portable Electric Tools
(See Hand-Tools,
No. 36)**

**51.   Power Transmission, (X_i) ☐   ☐
Mechanical**

a. Are all belts, pulleys, chains,
flywheels, shafting, and shaft
projections or other rotating or
reciprocating parts within 7 ft
of the floor of working platform effectively guarded?

$S_{i,j}$ _____

b. Are belts, pulleys, and shafting
located in rooms used exclu-
sively for power transmission
apparatus not guarded only
when the following requirements are met?
● The basement, tower, or room
occupied by transmission
equipment is locked against
unauthorized entrance.
● The vertical clearance in passageways between the floor

and power transmission
beams, ceiling, or any other
objects is not less than 5 ft.
- The intensity of illumination
  conforms to the required
  standards.
- The footing is dry, firm, and
  level.
- The route followed by the
  oiler is protected in such a
  manner as to prevent an accident.

$S_{i,j}$ _____

S_i _____

52. **Pressure Vessels, Port-** (X_i) ☐   ☐
    able Unfired

a.   Are all portable unfired pressure
     vessels designed and con-
     structed to meet the Standards
     of the American Society of
     Mechanical Engineers Boiler
     and Pressure Vessel Code, Sec-
     tion VIII or other equivalent
     codes?

$S_{i,j}$ _____

b.   Are portable unfired pressure
     vessels not built to code exam-
     ined quarterly by a competent
     person and subjected yearly to
     a hydrostatic pressure test of 1
     times the working pressure of
     the vessel? Are records of such
     examinations and tests main-
     tained?

$S_{i,j}$ _____

c.   Are relief valves on pressure
     vessels set to the safe working

pressure of the vessel, or to the
lowest safe working pressure of
the system, whichever is
lower?

$S_{i,j}$ _____

S_i _____

53.   Punch Presses (X_i) ☐   ☐

a.   Are "point-of-operation guards"
or properly applied and adjusted point-of-operation devices on every operation performed on a mechanical power
press provided? This requirement shall not apply when the
point-of-operation opening is $^1/_4$
in. or less.

$S_{i,j}$ _____

b.   Is a substantial guard placed
over the treadle of foot-
operated presses?

$S_{i,j}$ _____

c.   Do pedal counterweights, if provided on foot-operated presses,
have the path of the travel of
the weight enclosed?

$S_{i,j}$ _____

S_i _____

54.   Radiation (X_i) ☐   ☐

a.   Do the employers feel responsible for proper controls to prevent all employees from being
exposed to ionizing radiation in
excess of acceptable limits?

$S_{i,j}$ _____

b.   Is each radiation area conspicu-
      ously posted with appropriate
      signs?

                            $S_{i,j}$ _____

c.   Does employer maintain records
      of the radiation exposure of all
      employees for whom personnel
      monitoring is required?

                            $S_{i,j}$ _____
                            S_i _____

55.   Railings (X_i) ☐   ☐

a.   Does standard railing consist of
      top rail, intermediate rail, and
      posts and have a vertical height
      of 42 in. from upper surface of
      top rail to floor, platform, etc.?

                            $S_{i,j}$ _____

b.   Does railing for open-sided
      floors, platforms, and runways
      have a toeboard wherever per-
      sons can pass beneath the open
      side, there is moving machin-
      ery, or there is equipment that
      could be struck by falling ma-
      terials?

                            $S_{i,j}$ _____

c.   Are railings of such construction
      that the complete structure is
      capable of withstanding a load
      of at least 200 lb in any direc-
      tion on any point on the top
      rail?

                            $S_{i,j}$ _____

d.   Is the stair railing of construc-
      tion similar to a standard rail-

ing, but the vertical height not more than 34 in. nor less than 30 in. from upper surface of top rail to surface of tread in line with face or riser at forward edge of tread?

$S_{i,j}$ _____

S_i _____

56.   Rail Sweeps (See Cranes and Hoists, No. 15)

57.   Revolving Drums (X_i) ☐   ☐

a.   Are revolving drums, barrels, or containers guarded by an interlocked enclosure that prevents the drum (etc.) from revolving unless the guard enclosure is in place?

$S_{i,j}$ _____

S_i _____

58.   Saws, Band (Woodworking) (X_i) ☐   ☐

a.   Are all portions of band saw blades enclosed or guarded except for the working portion of the blade between the bottom of the guide rolls and the table?

$S_{i,j}$ _____

b.   Are band saw wheels fully encased? Is the outside periphery of the enclosure solid? Is the front and back either solid or wire mesh or perforated metal?

$S_{i,j}$ _____

S_i _____

59. Saws, Portable (X$_i$) ☐ ☐
Circular

a. Are portable power-driven circular saws having a blade diameter greater than 2 in. equipped
with guards above and below
the base plate or shoe? Do the
lower guards cover the saw to
the depth of the teeth, except
for the minimum arc required
to permit the base plate to be
titled for bevel cuts. Does it
automatically return to the covering position when the blade
is withdrawn from the work?

$S_{i,j}$ _____

S_i _____

60. Saws, Radial (Wood- (X$_i$) ☐ ☐
working)

a. Are radial saws constructed so
that the upper hood completely
encloses the upper portion of
the blade down to a point that
includes the end of the saw arbor? Is the upper hood constructed in such a manner and
of such a material that it pro-
tects the operator from flying
splinters, broken saw teeth,
etc. and deflects sawdust away
from the operator? Are the
sides of the lower exposed portion of the blade guarded to the
full diameter of the blade by a
device that automatically adjusts itself to the thickness of
the stock and remains in contact with stock being cut to

give the maximum protection
possible for the operation being
performed?

$$S_{i,j}$$ _____

b.   Do radial saws used for ripping
have non-kickback fingers or
dogs?

$$S_{i,j}$$ _____

c.   Are radial saws installed so that
the cutting head returns to the
starting position when released
by the operator?

$$S_{i,j}$$ _____
$$S_i$$ _____

**61.   Saws, Swing, or Slid-   (X_i)   ☐   ☐
ing Cutoff (Wood-
working)**

a.   Are swing or sliding cutoff saws
provided with a hood that completely encloses the upper half
of the saw?

$$S_{i,j}$$ _____

b.   Are limit stops provided to prevent swing or sliding type cutoff saws from extending beyond the front or back edges of
the table?

$$S_{i,j}$$ _____

c.   Is each swing or sliding cutoff
saw provided with an effective
device to return the saw automatically to the back of the table when released at any point
of its travel?

$$S_{i,j}$$ _____

d. Are inverted sawing or sliding cutoff saws provided with a hood that covers the part of the saw that protrudes above the top of the table or material being cut?

$S_{i,j}$ _____

S_i _____

62. Saws, Table (Wood- (X_i) ☐ ☐
working)

a. Do circular table saws have hoods over the portion of the saw above the table, so mounted that the hood automatically adjusts itself to the thickness and remains in contact with the material being cut?

$S_{i,j}$ _____

b. Do circular table saws have a spreader aligned with the blade, spaced no more than $\frac{1}{2}$ in. behind the largest blade mounted in the saw?

$S_{i,j}$ _____

c. Do circular table saws used for ripping have non-kickback fingers or dogs?

$S_{i,j}$ _____

d. Are freed rolls and blades of self-feed circular saws protected by a hood or guard to prevent the hands of the operator from coming in contact with the in-running rolls at any point?

$S_{i,j}$ _____

S_i _____

63.   Scaffolds   (X$_i$)   ☐   ☐

a.   Are all scaffolds and their sup-
       ports capable of supporting the
       load they are designed to carry
       with a factor of at least four?

                                S$_{i,j}$ _____

b.   Is all planking of scaffold grade,
       as recognized by grading rules
       for the species of wood used?
       Are the maximum permissible
       spans for 2 × 9 in. or wider
       planks according to the follow-
       ing table:

	Full thickness (undressed lumber)			Nominal thickness lumber	
Working load (psf)	25	50	75	25	50
Permissible span (ft)	10	8	6	8	6

       Does the maximum span for 1
       × 9 in. or wider plank of full
       thickness exceed 4 ft, with me-
       dium loading of 50 lb/f^2?

                                S$_{i,j}$ _____

c.   Do scaffold planks extend over
       their end supports not less than
       6 in. nor more than 18 in.?

                                S$_{i,j}$ _____

d.   Is scaffold planking overlapped
       a minimum of 12 in. or se-
       cured from movement?

                                S$_{i,j}$ _____

e.   Are railings and toeboards in-
       stalled on all open sides and
       ends of platforms more than 10
       ft above the floor except for

scaffolds covering an entire interior floor with no exposure to floor openings or needle-beam scaffolds used in structural iron work? Is there a screen with maximum $^1/_2$ in. openings between the toeboard and the top rail where persons are required to pass or work under the scaffold?

$S_{i,j}$ _____

S_i _____

64. **Spray-Finishing Operation** (X_i) ☐   ☐

a.   Is all spray finishing conducted in spray booths or spray rooms?

$S_{i,j}$ _____

b.   Are spray booths substantially constructed of steel, not thinner than No. 18 U.S. gauge, securely and rigidly supported, or of concrete or masonry; (aluminum or other substantial noncombustible material may be used for intermittent or low-volume spraying.) Are spray booths designed to sweep air currents toward the exhaust outlet?

$S_{i,j}$ _____

c.   Is there no open flame or spark-producing equipment in any spraying areas nor within 20 ft thereof, unless separated by a partition?

$S_{i,j}$ _____

d. Are electrical wiring and equipment not subject to deposits of combustible residues but located in a spraying area of an explosion-proof type approved for Class I, group D locations or for Class I, Division 1, Hazardous Locations? Is electrical wiring, motors, and other equipment outside of spraying area but within 20 ft of any spraying area, and not separated therefrom by partitions, produce sparks under normal operating conditions and otherwise conform to the provisions for Class I, Division 2, Hazardous Locations?

$S_{i,j}$ _____

e. Are all spraying areas provided with mechanical ventilation adequate to remove flammable vapors, mists, or powders to a safe location and to confine and control combustible residues so that life is not endangered?

$S_{i,j}$ _____

f. Are electric motors driving exhaust fans placed inside flammable materials spray booths or ducts? Are belts or pulleys within the booth or duct thoroughly enclosed?

$S_{i,j}$ _____

g. Is the quantity of flammable or combustible liquid kept in the vicinity of spraying operations the minimum required for operations? Does the quantity ordi-

narily not exceed a supply for
one day or one shift?

$S_{i,j}$ _____

h.   Are conspicuous "NO SMOK-
ING" signs posted at all flammable materials spraying areas
and storage rooms?

$S_{i,j}$ _____
S_i _____

65.   Stairs, Fixed Industrial (X_i) ☐   ☐

a.   Are standard railings provided
on the open sides of all exposed stairways? Are handrails
provided on at least one side of
closed stairways, preferably on
the right side descending?

$S_{i,j}$ _____

b.   Are stairs constructed so that
rise height and tread width is
uniform throughout?

$S_{i,j}$ _____

c.   Do fixed stairways have a minimum width of 22 in.?

$S_{i,j}$ _____
S_i _____

66.   Stationary Electrical (X_i) ☐   ☐
Devices

a.   Are all stationary electrically
powered equipment, tools, and
devices located within reach of
a person who can make contact
with any grounded surface or
object properly grounded?

$S_{i,j}$ _____
S_i _____

67.   **Storage** (X_i) ☐   ☐

a.   Is all storage stacked, blocked,
     interlocked, and limited in
     height so that it is secure
     against sliding or collapse?

                                 $S_{i,j}$ _____

b.   Are storage areas kept free from
     accumulation of materials that
     constitute hazards or pest har-
     borage? Is vegetation control
     exercised when necessary?

                                 $S_{i,j}$ _____

c.   Is sufficient safe clearance al-
     lowed for aisles, at loading
     docks, and through doorways
     where mechanical handling
     equipment is used?

                                 $S_{i,j}$ _____
                                 S_i _____

68.   **Tanks, Open-Surface** (X_i) ☐   ☐

a.   Is ventilation used to control po-
     tential exposures to employees
     adequate to reduce the concen-
     tration of the air contaminated
     to the degree that a hazard to
     employees does not exist?

                                 $S_{i,j}$ _____
                                 S_i _____

69.   **Toeboards** (X_i) ☐   ☐

a.   Are railings protecting floor
     openings, platforms, scaffolds,
     etc. equipped with toeboards
     whenever persons can pass be-
     neath the open side and there is

equipment that could be struck
by falling material?

$S_{i,j}$ _____

b.   Are standard toeboards at least 4
in. in height and of a substantial material, either solid or
open, with openings not exceeding 1 in. in greatest dimen-
sion?

$S_{i,j}$ _____
S_i _____

**70.   Toxic Vapors, Gases, (X_i) ☐   ☐
Mists, and Dusts**

a.   Is exposure to toxic vapors,
gases, mists, or dusts at a concentration above the threshold
limit values, contained or referred to in Safety and Health
Standards avoided?

$S_{i,j}$ _____

b.   To achieve compliance with
70.a, are administrative or engineering controls first determined and implemented whenever feasible? When such
controls are not feasible to
achieve full compliance, is protective equipment or any other
protective measure used to
keep the exposure of employees to air contaminants within
the limits prescribed? Is equipment and/or technical measures
used for this purpose approved
for each particular use by a
competent industrial hygienist

or other technically qualified
person?

$S_{i,j}$ _____

S_i _____

71.   Trash (X_i) ☐   ☐

a.   Is trash and rubbish collected
and removed in such a manner
as to avoid creating a menace
to health and as often as necessary to maintain good sanitary
conditions?

$S_{i,j}$ _____

S_i _____

72.   Washing Facilities (X_i) ☐   ☐

a.   Are adequate washing facilities
provided in every place of employment and maintained in a
sanitary condition? Is at least
one lavatory with adequate hot
and cold water provided for
every 10 employees up to 100
persons and one lavatory for
each 15 persons over 100?

$S_{i,j}$ _____

b.   Is a suitable cleaning agent, individual hand towels, or other
approved apparatus for drying
hands, and receptacles for dis-
posing of hand towels, provided at washing facilities?

$S_{i,j}$ _____

S_i _____

73.   Welding (X_i) ☐   ☐

a.   Is arc-welding equipment installed properly? Are workmen

designated to operate welding
equipment properly instructed
and qualified to operate it?

$S_{i,j}$ _____

b. Is mechanical ventilation provided when welding or cutting?
- Beryllium cadmium, lead,
zinc, or mercury
- Fluxes, metal coatings, or
other material containing fluorine compounds
- Where there is less than
10,000 ft³ per welder
- In confined spaces

$S_{i,j}$ _____

c. Is proper shielding and eye protection provided to prevent exposure of personnel from welding hazards?

$S_{i,j}$ _____

d. Are proper precautions (isolating
welding and cutting, removing
fire hazards from the vicinity,
providing a fire watch, etc.)
for fire prevention taken in
areas where welding or other
"hot work" is being done?

$S_{i,j}$ _____
S_i _____

74. Woodworking (X_i) ☐ ☐
Machinery

a. Is all woodworking machinery
such as table saws, swing
saws, radial saws, band saws,
jointers, tenoning machines,
boring and mortising machines,

shapers, planers, lathes, sanders, veneer cutters, and other miscellaneous woodworking machinery effectively guarded to protect the operator and other employees from hazards inherent to their operation?

$S_{i,j}$ _____

b.   Is a power-control device provided on each machine to make it possible for the operator to cut off the power from each machine, without leaving his position, at the point of operation?

$S_{i,j}$ _____

c.   Are power controls and operating controls located within easy reach of the operator while he is at his regular work location, making it unnecessary for him to reach over the cutter to make adjustments?

$S_{i,j}$ _____

d.   Are operating treadles protected against unexpected or accidental tripping?

$S_{i,j}$ _____
S_i _____

The material contained in this Check List is based on the suggestions contained in *Modern Safety & Health Technology* by R. DeReamer, John Wiley & Sons, 1980. (With permission.)

C.2 EVALUATION OF CONTRIBUTING CONDITIONS FOR ACCIDENTS

This checklist can be used for evaluating the major contributing factors for accidents for each department. These conditions are (1) supervisor's safety performance, (2) mental condition of worker, and (3) physical condition of worker. For each condition an appropriate score for each question should be circled.

1. Supervisor's Safety Performance

	Poor	Satisfactory	Good	Excellent
Has job hazard analysis been done for all the high and low hazard areas?	0	10	22	25
Have safety rules been applied to all equally and promptly?	0	2	8	10
Are regular safety training sessions held?	0	3	12	15
Have workers participated in safety improvement programs?	0	8	12	15
When workers are assigned to tasks, are the worker's capabilities matched with task requirements?	0	2	10	15
Have the recommendations of the safety inspections been incorporated when necessary?	0	3	14	20

Total value of circled numbers = _____ X. 4 rating = _____   (1)

2. Mental Condition of Worker

	Poor	Satisfactory	Good	Excellent
Had the supervisor regular safety contact with workers?	0	8	17	20
Had the worker been given adequate training before starting the job?	0	8	21	25
Was safety promotion and publicity done on regular basis?	0	4	11	15
Were safety meetings regularly held?	0	3	16	20
Did workers participate in safety improvements?	0	3	15	20

Total value of circled numbers = _____ X. 3 rating = _____   (2)

3. Physical Condition of Workers

	Poor	Satisfactory	Good	Excellent
Was the preplacement medical examination done?	0	7	20	25
Were periodic medical examinations done?	0	7	20	25
Were job placements properly done?	0	7	20	25
When workers changed jobs, were they physically examined?	0	7	20	25

Total value of circled numbers = _____ X. 3 rating = _____   (3)

Obtain overall Evaluation by adding Equations 1, 2, and 3.

APPENDIX D: SAFETY INFORMATION SYSTEM (SIS) DESCRIPTION AND THE USER MANUAL, AND SELECTED CODES FROM ANSI Z 16.2 (CHAPTER 6)

D.1   SIS DESCRIPTIONS AND THE USER MANUAL

SIS Computer Package

An implementation of the Safety Information System (SIS) has been accomplished by using an IBM PC. The safety data is organized into usable form by utilizing a data-base management system package called Database III (DB III or dbase III).

Database III allows creation of indexed files and manipulation of such files from stored programs. The indexing is required to minimize the time needed for sorting and searching for specific matches of data to meet user needs. The computer-to-user interface is via fill-in-the-blanks screen for soliciting input from the user and for reporting results to the user. Printed reports are also available for a permanent record.

The design of the program is tailored to take the best advantage of facilities provided by the data-base management package (DBIII); in addition, the programs are modular and generalized (i e , the same program is used for the generation of many compilation of reports). This is achieved by parameterizing the input to the program and then allowing the program to generate its own output based on the specific input at the time it is used.

DBIII Database Management System

The basic functions provided by this package are the following:

1. Creation of data files with text, numeric, and logical type fields
2. Creation of indexes into the above files; these indexes are then used to search and sort the records rapidly
3. Creation of fill-in-the-blanks screens which allow the passing of data from the screen to the program driving the screen
4. Creation of formulated reports from programs to the screen and to the printer

SIS Programming Philosophy

The SIS programs are written in DB III command language. This allows efficient interaction between the program and the screen and files.

The programs are modular, and a top-down hierarchy is established. These programs are even named such that each program position in the hierarchy is self-evident. All variable names are established according to the same naming conversation, i.e., each variable defined by a program starts with the program name.

Subroutines are used wherever possible to save processing time and space.

The size of each module has been kept small so that as each module is executed, the time to load it from disc to memory is the shortest possible.

SIS File Structure

The primary raw investigation data file is SIS. SIS contains codes for all possible items that can be collected about an investigation. Five files are described below.

The DESCOD file contains descriptions for each possible code value for each item in the SIS file. This is used whenever a report has to be generated. First, the codes are located and evaluated and then at the time of displaying the code, the description is obtained from DESCOD.

The DESFLD file contains the name of each data item field in the SIS file. This is used when field numbers or item numbers are available, but the SIS field name is required to access SIS. The field number is searched, and the corresponding description is obtained from the searched record.

The CNTR3-MI file is a temporary file which is used by CNTR5 program for storing the monthly injuries for statistical calculations and

---

*   The dBase-compatible computer program can be obtained free of charge from the authors, whose addresses are in the front of the book.

for the statistical comparison report. Once CALC3 program has used the data, it erases that data and the file is reused for the next report.

The CALC3-S1 file contains the injury frequencies means and standard deviations (SDs) for each department. These statistics can be displayed and/or updated every time a statistical report is requested. The file is also used for the reference purpose in comparing the injury means and SDs for a department or all departments.

The TDIST file contains the T-distribution values of significance levels 0.01 and 0.05 for spans of time in months starting from 5 months. This file is used for determining the level of accuracy required when a comparison report is requested.

SIS Programs

These are discussed below.

INITIAL — Initialize Index Program

This program is used for creating index files. It is a batch file that is executed once to create all the index files that are used in the SIS. The reason for keeping this file is to have a record of all the steps that were taken in building the file structures.

SISPROC-SIS Structure Procedure

This is a subroutine that is used to increment the case number. Every time the user wants to enter a new case, the SISPROC will increment the case number by taking into consideration the previous case number value. This special subroutine is used so that character type fields which contain numeric data have to be converted into numeric type and back to character type to perform any arithmetic.

SISI-SIS Main Program

This program allows the selection of functions or utilities that are available in SIS. It sits at the top of the hierarchy of SIS. It sets up DB III control paranthesis and activates SISPROC and closes the files when SIS is being exited.

DATA2 — Investigation Codes Entry and Update Program

This allows the entry of a new investigation case or the display of any previous case for reference or updating. It also processes the selection of updating the code description by calling DATA3 program. The case-number entered is searched for in the SIS file, and the entire investigation record is displayed. Any change can now be made by overwriting. New records are added by first entering "1" and then a fill-in-the-blank screen is provided for entering a code for the case to be added.

Code Description Entry Program

This is a program that allows the updating of the code descriptions in DESCOD file for the desired SIS code field. This means that new codes and their descriptions for all SIS field (items) can be added.

SLVR2 — Select Reporting Variable Program

This defines and initializes the report variables that will be processed in SLVR3 and SLVR4. In addition, SLVR2 provides a means of exiting out of the reporting mode or to request another report.

SLVR3 — Select Variable Report Options

This is used for entering report options, time spans, and variable choices, as well as the selection of department. It also displays the list of departments to choose from.

Furthermore, SLVR3 checks the time-spans for reasonability and checks the department code to make sure that such department exists. The variable selection is made by a number, and the number is converted to a field name by using DESFLD file.

SLVR4 — Set Up Variables and Titles for Reports

SLVR4 is called SLVR3 when it has obtained the report options. There on SLVR4 uses this information to assign search variables to search parameters also it assigns appropriate descriptions to titles from the DESFLD files.

When all the parameters have been assigned, the counting or calculation programs CNTR3 or CALC3 are called, depending on the report option selected.

It should be noted that this high degree of parameterization was necessary to limit the size of this package by using only a few generalized programs to process all combinations of report options.

CNTR3 — Counting Programs

This program uses the variable parameters set up by SLVR4 and the department code to dynamically build a compound index file. Parameter 1 (the primary variable) and parameter 2 are used as major and minor keys. The department is used as a super key. Furthermore, the data is used as a selection criteria for records to be included in the counts. The count is carried out by CNTR4 and CNTR5 for univariate and bivariate, respectively. Both the headings for bivariate reports and for the minor search variable are printed.

CNTR4 — Process Univariate Reports

This program processes univariate counting and prints minor and total counts. The incidence rate data collection is also accomplished here to be further processed and printed by INCD5. CNTR4 calls CNTR6 to generate heading for univariate reports and create the required index files. On return from CNTR6, the counts are printed. Furthermore, special processing is performed for the summary of investigation report.

CNTR5 — Process and Print Bivariate Reports

This program counts minor and major frequencies for the bivariate reports. Also, it assigns the monthly injury frequencies to be used by the CALC3 program. Furthermore, the program processes a special report for univariate frequencies by department. The proper alignment of data under column headings and line scrolling is accomplished. It is to be noted that the univariate reports are quite complex and require sophisticated screen control to ensure that the counts will not be written over.

CNTR6 — Calculate Indexes and Headings

This program prints headings and creates appropriate index files for univariate and incidence rate reports.

INCD5 — Incidence Rate Processing

This program allows the selection of additional information to be used in the report in question. It uses information (such as number of worker hours) to calculate incidence rates for a department or for all departments. The final report headings and incidence rate results and totals are printed. INCD5 maintains several cumulative sums to allow comprehensive calculations of injury and lost day/cost severities.

CALC3 — Calculate and Compare Injury Statistics

This program uses the CNTR3-M1 file to calculate means and standard deviations of monthly injury frequencies and saves them in CALC3-S1 file for subsequent comparison purpose. It also allows the comparison of the monthly injury frequencies for a specified span of time for a department or all departments to the overall mean and standard deviation maintained in CALC3-S1 file. Furthermore, it uses the TDIST file to assign T-distribution values by span of time and for 0.01 and 0.05 significance levels.

The report results are printed after the presentation of statistical calculations.

LIST2 — Listing Initiation Program

This program defines and initializes the listing variables to be used by subsequent programs. It creates the option selection screen and asks for the time spans. Furthermore, it displays departments and their codes for the selection of the user and processes the listing selection variables. Subsequently, the program submits this information to LIST3 for further processing.

LIST3 — Selection Processing

This program processes option selections made in LIST2 and allows the selection of code value range within variables. Also, it checks for validation of code values entered and then submits this information to LIST4 for searching and processing.

LIST4 — Bivariate Listing Processing

This program uses variable parameters set up by LIST2 and LIST3 and the department code to immediately build a compound index file. Parameter 1 (the primary variable) and parameter 2 are used as major and minor searching keys. The department code is used as a super key. The data is used as a checking criteria for records to be listed. LIST4 does the bivariate listing as well as the univariate listing by department. It calls LIST5 for univariate listing and LIST6 to print the selected records.

LIST5 — Univariate Listing Processing

This program uses the parameters set up by LIST2 and the department code to build the compound index file. This parameter is used as the searching key with the department code being used as the super key. LIST5 processes the univariate listing functions and then calls LIST6 to print the selected records.

LIST6 — Records Printing

This program prints the records selected in LIST4 or LIST5. In addition, it accesses the codes description file DESCOD and the field name file DESFLD to print the records in an understandable form.

Furthermore, it associates the code values with their description in the listing for reference purposes.

D.1.1   User Manual

The SIS is a self-explanatory user-friendly system, and the user will find no difficulties in operating it. However, this user manual contains a brief instruction to be used as a reference for the user and to fulfill any inquiry on how to operate the system.

1. Operation Requirements
   In order to use the SIS on an IBM PC, Operating System Disk "DOS", dbase III system disc, and the SIS disc must be provided. If reports are to be printed, the EPSON FX 80 or equivalent printer is required.

2. Starting the System
   1. Insert the system operating disk DOS in Drive A and boot the system.
   2. When the prompt (A) appears on the screen, take out DOS and insert dbase III disk in Drive A and SIS disk in Drive B. Write dbase and hit return (—-). First, the dbase screen will appear, then the SIS screen will follow.

3. Safety Information System
   The main menu of SIS will appear on the screen asking the user to select one of the facilities provided by SIS. The user shall enter:
   1. for data entry
   2. for data listing
   3. for reports
   4. to quit the system
   Having entered the selected number, the system will take you to the corresponding second menu.

Data Entry
If "1" is entered in the main menu, the data entry menu will appear on the screen asking the user to select the desired function. The choices are (enter the number)

1. for entering new investigation case
2. for entering code description

3.   for showing a certain investigation case
4.   to go back to the main menu

1.   Entering New Investigation Case
     If "1" is entered in the data entry menu, a new screen will appear with the new case number assigned and the data base fields shown empty. The user enters the new case information and with the last bit of information (investigator code) the system will take you back to the data entry screen.

2.   Entering Codes Description
     If "2" is entered in the data entry menu, a new screen will appear with the data items for which the user can update the codes descriptions. Enter the variable code and a new screen appears. The codes and descriptions will be listed one after another, and the user can update these descriptions. With the last description entered, the system will take you back to the data entry menu.

3.   Showing a Certain Investigation Case
     If "3" is entered, another message will appear on the screen asking for the case number. Enter the desired case number; then the system will ask the user if this case information is to be printed or displayed on the screen. Having chosen that, the data will be either printed or displayed, then press any key to go back to the data entry menu.

4.   Quitting the Data Entry Menu
     If Q is entered, the system will take you back to the main menu.

Data List
If "2" is entered in the main menu, the data list menu will appear on the screen, asking the user to select the selection criteria by number.

1.   for univariate listing by department
2.   for univariate listing
3.   for bivariate listing
Q.   for going back to the main menu

After the selection, messages will appear on the screen asking the user to enter the beginning year, month, and day and the ending year, month, and day and to specify the time span period for which the records are to be listed.

1. Univariate Listing by Department

    The variables will be listed for selection, and the user shall enter the variable code. Then the code values of that variable are to be specified. A new screen will appear with a message asking the user if he/she wants to print or display the data or quit. Then, accordingly, the data will be printed or displayed for the first record. Later on, the same message will appear again to either print or display the second record or quit. If Q is entered, the system will take you back to the data listing menu.

2. Univariate Listing

    After the data message, the system will display the depart-ments and their codes to select the department. The rest will be the same as in the case of univariate listing by department.

3. Bivariate Listing

    This is the same as for the univariate listing, but another variable is to be entered and, as a consequence, the code values of that variable will be entered.

Q. Going Back to the Main Menu

    If Q is entered, the system will take you back to the main menu.

Reports

If "3" is entered in the main menu, the report menu will appear on the screen. There are 7 different kinds of reports and the quitting choice:

1. investigation summary report
2. univariate by department report
3. univariate distribution report
4. bivariate distribution report
5. statistical calculation report
6. monthly incidence rates report
7. statistical comparison report
Q. going back to the main menu

If any report is selected, the system will ask the user to enter the beginning and ending dates of the time span for which the report is to be generated, except for reports 5, 6, and 7, where the beginning and ending days are not required. Subsequently, the departments will be displayed to select the desired department, except for Report 2 where department code is not required. For all reports, before displaying the report, the system will ask the user if he/she wants to print or display.

1. Investigation Summary Report

   After entering the date and department, the message of printing or displaying will appear. Then, accordingly, the report will be either printed or displayed. At a later stage the system will send a message on the screen asking the user to press any key; to continue pressing any key will take you back to the report menu.

2. Univariate By Department Report

   After entering the dates, the variables will be displayed on the screen and the user should enter the desired variable. Then he/she should proceed as in Report 1.

3. Univariate Distribution Report

   This report is the same as Report 2.

4. Bivariate Distribution Report

   In this report, the user shall enter two variables; the rest is just like Reports 2 and 3.

5. Statistical Calculation Report

   This report will proceed in the same manner as in Report 1.

6. Monthly Incidence Rate Report

   This report will proceed in the same manner as in Report 5, but inside the report, the system will ask the user to enter the number of work days, average workers per shift, number of shifts, and the duration of shift for each month. When all months are completed, the totals will be printed. Then press any key to go back to the report menu.

7. Statistical Comparison Report

   This report is the same as Report 5 except that the user will be asked to enter the significance level.

Q. Going Back to the Main Menu

   If Q is entered in the report menu, the system will take you back to the main menu.

Q. Quitting from the SIS.

   If Q is entered in the main menu, the SIS screen will appear and then you are out of the SIS and the Database III.

D.2 SELECTED CODES FROM ANSI Z 16.2–1962 (R1969)

These codes are intended to serve as a reference list for items 18, 19, . . . , and 23, in the data base.

1. Source of Injury Classification

Code used in dBASE	Standard code	Description
0001	0100	Air pressure (abnormal environmental)
0002	0200	Animals, insects, birds, reptiles (live)
0003	0300	Animal products (not food)
0004	0400	Bodily motion (no lifting, pulling, pushing, etc. See Rule 3.3.2.3)
0005	0500	Boilers, pressure vessels
0006	0600	Boxes, barrels, containers, packages (empty or full)
0007	0700	Buildings and structures (not floors, working surfaces, or walkways; see working surfaces)
0008	0800	Ceramic items, NEC*
0009	0900	Chemicals, chemical compounds (solids, liquids, gases)
0010	1000	Clothing, apparel, shoes
0011	1100	Coal and petroleum products
0012	1200	Cold (atmospheric, environmental)
0013	1300	Conveyors
0014	1400	Drugs and medicines
0015	1500	Electric apparatus
0016	1700	Flame, fire, smoke
0017	1800	Food products (including animal foods)
0018	1900	Furniture, fixtures, furnishings (not fixed parts of buildings or structures)
0019	2000	Glass items, NEC (glassware, glass fibers, sheets, etc. — not bottles, jars, flasks, or glass cloth)
0020	2200	Hand tools, not powered
0021	2300	Hand tools, powered
0022	2400	Heat, atmospheric, environmental (not hot objects or substances)
0023	2500	Heating equipment, NEC (furnaces, retorts, space heaters, stoves, ranges, etc. — not electric)
0024	2600	Hoisting apparatus
0025	2700	Infectious and parasitic agents, NEC (bacteria, fungi, parasitic organisms, viruses, etc. — not chemicals, drugs, prepared medicinal items, or food infestations)
0026	2800	Ladders (fixed or portable)
0027	2900	Liquids, NEC
0028	3000	Machines
0029	4000	Mechanical power transmission apparatus *Note:* Transmission equipment includes all mechanical means of transmitting power from a prime mover up to but not including a shaft, or any pulleys or gears on the shaft, the bearings of which form an integral part of a machine. Directly connected prime movers are defined as having no transmission apparatus.
0030	4100	Metal items, NEC (plates, rods, wire, shapes, nuts, bolts, nails, etc. — includes molten metal, ingots, and melting scrap, but not ores or other raw materials)

* NEC = Not Elsewhere Identified.

Code used in dBASE	Standard code	Description
0031	4200	Mineral items, metallic, NEC (products of mining, excavating, landslides, etc. such as dirt, clay, sand, gravel, stone, etc.)
0032	4400	Noise
0033	4500	Paper and pulp items, NEC
0034	4600	Particles (unidentified)
0035	4700	Plants, trees, vegetation (in natural or unprocessed condition — does not include threshed grains, harvested fruits, limbed logs, etc.)
0036	4800	Plastic items, NEC (powder, sheets, rods, shapes, etc., but not uncombined chemicals or components used in plastic manufacturing)
0037	4900	Pumps and prime movers
0038	5000	Radiating substances and equipment (use this code only in cases of radiation injuries)
0039	5100	Soaps, detergents, cleaning compounds
0040	5200	Silica
0041	5300	Scrap, debris, waste materials, etc.
0042	5400	Steam
0043	5500	Textile items, NEC (includes animal fibers after first scouring and cleaning, vegetable and synthetic fibers — except glass — yarn, thread, yard goods, felts, and textile products)
0044	5600	Vehicles (see Rule 3.3.2.4 regarding injuries experienced while occupying a vehicle)
0045	5700	Wood items, NEC (logs, lumber, slabs, chips, and wood products)
0046	5800	Working surfaces (surfaces in use as supports for people)
0047	8800	Miscellaneous, NEC
0048	9800	Unknown, unidentified (other than particles)

2. Accident Type Classification

This category includes accidents other than motor-vehicle or public transportation accidents.

Code used in dBASE	Standard code	Description
	010	**Struck against**
0001	011	Stationary object
0002	012	Moving objects
	020	**Struck by**
0003	021	Falling object
0004	022	Flying object
0005	029	Struck by, NEC*
	030	**Fall from elevation**
0006	031	From scaffolds, walkways, platforms, etc.
0007	032	From ladders
0008	033	From piled materials
0009	034	From vehicles
0010	035	On stairs
0011	036	Into shafts, excavations, floor openings, etc. (from edge of opening)

* NEC = Not Elsewhere Identified.

Code used in dBASE	Standard code	Description
0012	039	Fall to lower level, NEC
	050	**Fall on same level**
0013	051	Fall to the walkway or working surface
0014	052	Fall onto or against objects
0015	059	Fall on same level, NEC
	060	**Caught in, under, or between**
0016	061	Inrunning or meshing objects
0017	062	A moving and a stationary object
0018	063	Two or more moving (not meshing) objects
0019	064	Collapsing materials (slides of earth, collapse of buildings, etc.)
0020	069	Caught in, under, or between, NEC
	080	**Rubbed or abraded**
0021	081	By leaning, kneeling, or sitting on objects (not vibrating)
0022	082	By objects being handled (not vibrating)
0023	083	By vibrating objects
0024	084	By foreign matter in eyes
0025	085	By repetition of pressure
0026	089	Rubbed or abraded, NEC
	100	**Bodily reaction**
0027	101	From involuntary motions
0028	102	From voluntary motions
	120	**Overexertion**
0029	121	In lifting objects
0030	122	In pulling or pushing objects
0031	123	In wielding or throwing objects
0032	129	Overexertion, NEC
	130	**Contact with electric current**
	150	**Contact with temperature extremes**
0033	151	General heat — atmosphere or environment
0034	152	General cold — atmosphere or environment
0035	153	Hot objects or substances
0037	154	Cold objects or substances
	180	**Contact with radiations, caustics, toxic and noxious substances**
0038	181	By inhalation
0039	182	By ingestion
0040	183	By absorption
0041	189	NEC
	200	**Public transportation accidents** (code for type of vehicle in which injured was a passenger)
0042	201	Aircraft accident
0043	203	Bus accident
0044	205	Ship or boat accident
0045	207	Streetcar or subway accident
0046	209	Taxi accident
0047	211	Train accident
0048	298	Public vehicle accident, NEC
	300	**Motor-vehicle accidents** (code in terms of the event affecting or involving the vehicle in which the injured was an occupant. If more than one of the listed events occurred, code for the first event in the sequence)
	310	Collision or sideswipe with another vehicle — both vehicles in motion
0049	311	With an oncoming vehicle on same road, street, or trafficway
0050	312	With a vehicle moving in same direction on same road, street, or trafficway

Code used in dBASE	Standard code	Description
0051	313	With a vehicle moving in an intersecting trafficway
	320	Collision or sideswipe with a standing vehicle or stationary object
0052	321	Running into or sideswiping a standing vehicle or object in the roadway
0053	322	Running into or sideswiping a standing vehicle or object at side of road (not in trafficway)
0054	323	Struck by another vehicle while standing in roadway
0055	324	Struck by another vehicle while standing off the roadway
	330	Noncollision accidents
0056	331	Overturned
0057	332	Ran off roadway (out of control)
0058	333	Sudden stop or start (throwing occupants out of, or against interior parts of the vehicle; or throwing contents of vehicle against occupants)
0059	338	Other noncollision accidents
0060	899	**Accident type, NEC**
0061	999	**Unclassified, insufficient data**

3. Hazardous Condition Classification

Code used in dBASE	Standard code	Description
	000	**Defects of agencies** (i.e., undesired and unintended characteristics, generally the opposite of the desirable and proper characteristics, such as being dull when it should be sharp. Do not classify an intended and necessary characteristic of an agency as a defect. For example: a knife is expected to be sharp and is not defective because it has this characteristic.)
0001	001	Composed of unsuitable materials
0002	005	Dull
0003	010	Improperly compounded, constructed, or assembled
0004	015	Improperly designed
0005	020	Rough
0006	025	Sharp
0007	030	Slippery
0008	035	Worn, cracked, frayed, broken, etc.
0009	099	Other defects, NEC*
	100	**Dress or apparel hazards** *Note:* Name this hazardous condition if it, in fact, contributed to the occurrence of the accident even though the condition was created by the injured employee's own choice or unsafe act.
0010	110	Lack of necessary personal protection equipment *Note:* Name this hazard only when the personal protective equipment constitutes an essential element in the safe performance of the activity. Does not apply when the use of the protective equipment would merely have minimized the injury without preventing the accident.
0011	113	Improper or inadequate clothing
0012	199	Dress or apparel hazards, NEC

* NEC = Not Elswhere Identified.

Code used in dBASE	Standard code	Description
	200	**Environmental hazards,** NEC
		Note: These are general hazards of the workplace which commonly affect everyone in the area regardless of his assignment. They should be named as the accident cause only when none of the other more specific hazardous condition designations apply.
0013	205	Excessive noise
0014	210	Inadequate aisle space, exits, etc.
0015	220	Inadequate clearance (for moving objects or persons)
0016	230	Inadequate traffic control (on employers' premises only — refers to maintenance of traffic lanes; elimination of blind corners, etc.; control of speeding; direction of traffic away from danger points, etc.)
0017	240	Inadequate ventilation (general — not due to defective equipment)
0018	250	Insufficient workspace
0019	260	Improper illumination (insufficient light for the operation, glare, etc.)
0020	299	Environmental hazards, NEC
	300	**Hazardous methods or procedures**
		(Caution should be observed in the application of this classification, particularly to avoid its becoming a "catch-all" for cases which cannot be assigned to other specific classifications because of inadequate information. It is not intended that an activity should be classified as a hazardous procedure simply because an injury or injuries occurred in the course of that activity. A hazardous method or procedure in this context is usually a deviation from the normal and generally accepted safe procedures commonly applied in industrial operations. In some respects, this classification parallels the unsafe act classification. The distinguishing characteristic is that the procedures classified here were planned, directed, or conducted by supervision.)
0021	310	Use of inherently hazardous (not defective) material or equipment
0022	320	Use of inherently hazardous methods or procedures
0023	330	Use of inadequate (not defective) or improper tools or equipment
0024	340	Inadequate help for heavy lifting, etc.
0025	350	Improper assignment of personnel (i.e., disregard of physical limitations, skill, etc.)
0026	399	Hazardous methods or procedures, NEC
	400	**Placement hazards** (materials, equipment, etc. — not persons)
0027	410	Improperly piled (refers to manner of piling)
0028	420	Improperly placed (refers to position occupied)
0029	530	Inadequately secured against undesired motion (not unstable piling)
	500	**Inadequately guarded**
0030	510	Unguarded (mechanical or physical hazards — not electrical or radiation hazards)
0031	520	Inadequately guarded (mechanical or physical hazards — not electrical or radiation hazards)
0032	530	Lack of or inadequate shoring in mining, excavating, construction, etc.
0033	540	Underground (electrical)
0034	550	Uninsulated (electrical)

Code used in dBASE	Standard code	Description
0035	560	Uncovered connections, switches, etc. (electrical)
0036	570	Unshielded (radiation)
0037	580	Inadequately shielded (radiation)
0038	590	Unlabeled or inadequately labeled materials
0039	599	Inadequately guarded, NEC
	600	**Hazards of outside work enviroments — other than public hazards** (encountered while working in or on premises not controlled by the employer and not arising from the activities of the injured or co-employees or from the tools, materials, or equipment used in those activities)
0040	610	Defective premises of others
0041	620	Defective materials or equipment of others
0042	630	Other hazards associated with the property or operations of others
0043	640	Natural hazards (i.e., hazards of irregular and unstable terrain; exposure to the elements, wild animals, etc.; encountered in open country operations but not in cleared or regularly designated work areas)
	700	**Public hazards** (encountered on public places away from employers' premises)
0044	710	Public transportation hazards (encountered while a passenger is on a public carrier)
0045	720	Traffic hazards (encountered on public streets, roads, or highways)
0046	780	Other public hazards (other hazards of public places to which the general public is also exposed)
0047	980	**Hazardous conditions,** NEC
0048	990	**Undetermined — insufficient information**
0049	999	**No hazardous condition**

4. Nature of Injury Classification

Code used in dBASE	Standard code	Description
0100	100	Amputation or enucleation
0110	110	Asphyxia, strangulation, drowning
0120	120	Burn or scald (heat) — the effect of contact with hot substances. Includes electric burns, but not electric shock. Does not include chemical burns, effects of radiation, sunburn, systemic disability such as heat stroke, friction burns, etc.
0130	130	Burn (chemical) — tissue damage resulting from the corrosive action of chemicals, chemical compounds, fumes, etc. (acids, alkalies)
0140	140	Concussion — brain, cerebral
0150	150	Contagious or infectious disease — anthrax, brucellosis, tuberculosis, etc.
0160	160	Contusion, crushing, bruise — intact skin surface
0170	170	Cut, laceration, puncture — open wound
0180	180	Dermatitis — rash, skin or tissue inflammation, including boils, etc. Generally resulting from direct contact with irritants or sensitizing chemicals such as drugs, oils, biologic reagents, plants, woods, or metals, which may be in the form of solids, pastes, liquids, or vapors and which may be contacted in the pure state or in compounds or in combination with other materials. Does not include skin or tissue damage resulting from

Code used in dBASE	Standard code	Description
		corrosive action of chemicals, burns from contact with hot substances, effects of exposure to radiation, effects of exposure to low temperatures, or inflammation or irritation resulting from friction or impact
0190	190	Dislocation
0200	200	Electric shock, electrocution
0210	210	Fracture
0220	220	Freezing, frostbite, and other effects of exposure to low temperature
0230	230	Hearing loss, or impairment (a separate injury, not the sequelae of another injury)
0240	240	Heat stroke, sunstroke, heat cramps, heat exhaustion, and other effects of environmental heat. Does not include sunburn or other effects of radiation
0250	250	Hernia, rupture — includes both inguinal and noninguinal hernias
0260	260	Inflammation or irritation of joints, tendons, or muscles — includes bursitis, synovitis, tenosynovitis, etc. Does not include strains, sprains, or dislocation of muscles or tendons, or their aftereffects.
0270	270	Poisoning, systemic — a systemic morbid condition resulting from the inhalation, ingestion, or skin absorption of a toxic substance affecting the functioning of the metabolic system, the nervous system, the circulatory system, the digestive system, the respiratory system, the excretory system, the musculoskeletal system, etc. Includes chemical or drug poisoning, metal poisoning, organic diseases, and venomous reptile and insect bites. Does not include effects of radiation, pneumoconiosis, corrosive effects of chemicals, skin-surface irritations; septicemia or infected wounds
0280	280	Pneumoconiosis — includes anthrocosis, asbestosis, silicosis, etc.
0290	290	Radiation effects — sunburn and all forms of damage to tissue, bones, or body fluids produced by exposure to radiations
0300	300	Scratches, abrasions (superficial wounds)
0310	310	Sprains, strains
0400	400	Multiple injuries
0990	990	Occupational disease, NEC*
0995	995	Other injury, NEC
0999	999	Unclassified, not determined

5. Part of Body Affected Classification

Code used in dBASE	Standard code	Description
	100	**Head**
0110	110	Brain
	120	Ear(s)
0121	121	Ear(s) external
0124	124	Ear(s) internal (include hearing)
0130	130	Eye(s) (include optic nerves and vision)
	140	Face
0141	141	Jaw (include chin)

* NEC = Not Elsewhere Identified.

Code used in dBASE	Standard code	Description
0144	144	Mouth (include lips, teeth, tongue, throat, and taste)
0146	146	Nose (include nasal passages, sinus, and sense of smell)
0148	148	Face, multiple parts (any combination of above parts)
0149	149	Face, NEC*
0150	150	Scalp
0160	160	Skull
0198	198	Head, multiple (any combination of above parts)
0199	199	Head, NEC
	200	**Neck**
	300	**Upper extremities**
	310	Arm(s) (above wrist)
0311	311	Upper arm
0313	313	Elbow
0315	315	Forearm
0318	318	Arm, multiple (any combination of above parts)
0319	319	Arm, NEC
0320	320	Wrist
0330	330	Hand (not wrist or fingers)
0340	340	Finger(s)
0398	398	Upper extremities, multiple (any combination of above parts)
0399	399	Upper extremities, NEC
	400	**Trunk**
0410	410	Abdomen (include internal organs)
0420	420	Back (include back muscles, spine, and spinal cord)
0430	430	Chest (include ribs, breast bone, and internal organs of the chest)
0440	440	Hips (include pelvis, pelvic organs, and buttocks)
0450	450	Shoulder(s)
0498	498	Trunk, multiple (any combination of above parts)
0499	499	Trunk, NEC
	500	**Lower extremities**
	510	Leg(s) (above ankle)
0511	511	Thigh
0513	513	Knee
0515	515	Lower leg
0518	518	Leg, multiple (any combination of above parts)
0519	519	Leg, NEC
0520	520	Ankle
0530	530	Foot (not ankle or toes)
0540	540	Toe(s)
0598	598	Lower extremities, multiple (any combination of above parts)
0599	599	Lower extremities, NEC
0700	700	**Multiple parts** (applies when more than one major body part has been affected, such as an arm and a leg)
	800	**Body system** (applies when the functioning of an entire body system has been affected without specific injury to any other part, as in the case of poisoning, corrosive action affecting internal organs, damage to nerve centers, etc. Does not apply when the systemic damage results from an external injury affecting an external part such as a back injury which includes damage to the nerves of the spinal cord.)
0801	801	Circulatory system (heart, blood, arteries, veins, etc.)
0810	810	Digestive system
0820	820	Excretory system (kidneys, bladder, intestines, etc.)
0830	830	Musculo-skeletal system (bones, joints, tendons, muscles, etc.)

* NEC = Not Elsewhere Identified.

Code used in dBASE	Standard code	Description
0840	840	Nervous system
0850	850	Respiratory system (lungs, etc.)
0880	880	Other body systems
0900	900	**Body parts,** NEC
0999	999	**Unclassified** (insufficient information to identify part affected)

6. Unsafe Act Classification

Code used in dBASE	Standard code	Description
	050	**Cleaning, oiling, adjusting, or repairing of moving, electrically energized, or pressurized equipment** (Does not include actions directed by supervision.)
0051	051	Caulking, packing, etc. of equipment under pressure (pressure vessels, valves, joints, pipes, fittings, etc.)
0052	052	Cleaning, oiling, adjusting, etc. of moving equipment
0056	056	Welding, repairing, etc. of tanks, containers, or equipment without supervisory clearance in respect to the presence of dangerous vapors, chemicals, etc.
0057	057	Working on electrically charged equipment (motors, generators, lines, etc.)
0059	059	NEC*
0100	100	**Failure to use available personal protective equipment** (goggles, gloves, masks, aprons, hats, lifelines, shoes, etc.)
0150	150	**Failure to wear safe personal attire** (wearing high heels, loose hair, long sleeves, loose clothing, etc.)
	200	**Failure to secure or warn**
0201	201	Failure to lock, block, or secure vehicles, switches, valves, press rams, other tools, materials, and equipment against unexpected motion, flow of electric current, steam, etc.
0202	202	Failure to shut off equipment not in use
0203	203	Failure to place warning signs, signals, tags, etc.
0205	205	Releasing or moving loads, etc. without giving adequate warning
0207	207	Starting or stopping plant vehicles or equipment without giving adequate warning
0209	209	NEC
0250	250	**Horseplay** (distracting, teasing, abusing, startling, quarreling, practical joking, throwing material, showing off, etc.)
	300	**Improper use of equipment**
0301	301	Use of material or equipment in a manner for which it was not intended
0305	305	Overloading (vehicles, scaffolds, etc.)
0309	309	NEC
	350	**Improper use of hands or body parts**
0353	353	Gripping objects insecurely
0355	355	Taking wrong hold of objects
0356	356	Using hands instead of hand tools (to feed, clean, adjust, repair, etc.)
0359	359	NEC
0400	400	**Inattention to footing or surroundings**
	450	**Making safety devices inoperative**
0452	452	Blocking, plugging, tying, etc. of safety devices

* NEC = Not Elsewhere Identified.

Code used in dBASE	Standard code	Description
0453	453	Disconnecting or removing safety devices
0454	454	Misadjusting safety devices
0456	456	Replacing safety devices with those of improper capacity (e.g., higher amperage electric fuse, low capacity safety valves, etc.)
0459	459	NEC
	500	**Operating or working at unsafe speed**
0502	502	Feeding or supplying too rapidly
0503	503	Jumping from elevations (vehicles, platforms, etc.)
0505	505	Operating plant vehicles at unsafe speed
0506	506	Running
0508	508	Throwing material instead of carrying or passing it
0509	509	NEC
	550	**Taking unsafe position or posture**
0552	552	Entering tanks, bins, or other enclosed spaces without proper surpervisory clearance
0555	555	Riding in unsafe position (e.g., on platforms, tailboards, on running boards of vehicles; on forks or lift truck; on hook of cranes; etc.)
0556	556	Unnecessary exposure to undersuspended loads
0557	557	Unnecessary exposure to swinging loads
0558	558	Unnecessary exposure to moving materials or equipment
0559	559	NEC
	600	**Driving errors** (by vehicle operator on public roadways)
0601	601	Driving too fast or too slowly
0602	602	Entering or leaving vehicle on traffic side
0603	603	Failure to signal when stopping, turning, backing
0604	604	Failure to yield right of way
0605	605	Failure to obey traffic control signs or signals
0606	606	Following too closely
0607	607	Improper passing
0608	608	Turn from wrong lane
0609	609	NEC
	650	**Unsafe placing, mixing, combining, etc.**
0653	653	Injecting, mixing, or combining one substance with another so that explosion, fire, or other hazard is created (e.g., injecting cold water into hot boiler, pouring water into acid, etc.)
0655	655	Unsafe placing of vehicles or material moving equipment (i.e., parking, placing, stopping, or leaving vehicles, elevators, or conveying apparatus in unsafe position for loading or unloading)
0657	657	Unsafe placement of materials, tools, scrap, etc. (i.e., so as to create tripping, bumping, slipping hazards, etc.)
0659	659	NEC
0750	750	**Using unsafe equipment** (e.g., equipment tagged as defective or obviously defective. Does not include the use of inherently hazardous material for its intended purpose unless it was obviously defective. Does not include defective material or equipment when the defect was hidden and not obvious to the user.)
0900	900	**Unsafe act,** NEC
0993	993	**No unsafe act**
0999	999	**Unclassified — inadequate data**

INDEX

A

Abilities, 15
Abrasive-blasting respirators, 117
Abrasive blasting safety, 116–117
Abrasive grinding, 117–118
Abrasive wheels, 117–118
Accident level, 14
Accidents
  causes of, 75–76
    classification of, 61, 63
    research on, 11–12
  contributing causes of, 75, 80–82
  contributing conditions of, 162–163
  control chart of, 55–58
  cost measurements for, 25–30
  definition of, 8
  factors in, 14
  incidence rates of, 92–93
  monthly distribution of, 36
  by organization, 88
  prevention of, 11, 12
  rates of, 21
  record of, 35
  statistical analysis of, 90–94
  type classification of, 176–178
Accuracy, 15
Acetylene cylinders, 126
ADC, see Average Days Charged
Air receivers, compressed, 118–119
Air tools, 119
Aisles
  safety checklist for, 119–120
  in storage areas, 159
Allen, Zechariah, 1–2
American National Standards Institute
  (ANSI) indices, 17–20
American Public Health Association,
  2
American Society of Mechanical
  Engineers Boiler and Pressure
  Vessel Code, Section VIII, 148
ANSI A11.1 Standard, 143

ANSI Z 16.2–1962 (R1969) Standard
  codes, 175–184
ANSI Z 16.1–1967 (R1973) Standard
  for Recording and Measuring
  Injuries, 17–20
ANSI Z89.1–1967 Standard, 138
Arc-welding equipment, 160
Attitudes, worker, 81
Authority, operating without, 109
Automation
  equipment safety and, 101–102
  improved safety with, 98–102
  level of, 96–98
  worker's performance and, 95–98
Average Days Charged, 17, 19–20,
  23–24

B

Band saws, 151–152
Bauer, George, 1
Belt sanding machines, 120
Beryllium welding, 161
Bivariate distribution, 92
Blast-cleaning enclosures, 116
Blasting agents, 130
Body, emergency flushing of, 129
Boilers, 120–121
Books, safety-related, 4–6
Bureau of Labor Statistics (BLS)/
  OSHA rates, 20–21

C

Cables, 121–122
Cadmium welding, 161
Calendars, 121
Carbon tetrachloride, 122
Carrying, unsafe, 111
Chains, 121–122
Child Protection and Toy Safety Act,
  2
Chip guards, 122
Chlorinated hydrocarbons, 122–123

Circular saws, 151
Code, 8
Combining, unsafe, 111
Combustible liquids
  dip tanks for, 126–127
  drains for, 128
  incidental to principal business,
    132–134
Compressed air, 123
Compressed gas cylinders, 125–126
Cone pulleys, 123
Confidence interval, 8
Confidence level, 9
Consumer Product Safety Act, 2
Control chart, 55–58
  definition of, 8
  safety behavior, 71–72
Control errors, 99
Control limit, 8
Conveyors, 124
Cost measurements, accident, 25–30
Cranes, 124–125
Crowding, 111
Cumulative probability distribution,
    38–39
Cylinders, 125–126

D

Data base, 88–90, 165–171
Data sources, 7
Database III system, 88, 165–171
Davy, Humphrey, 1
dBase III package, see Database III
    system
De Reamer, R., 75
Defense Department, safety
    documents of, 2–3
DELPHI Method, 83–84
Dip tanks, 126–127
Direct manual work, 98
Disabling injuries
  recorded over time, 56
  trends in, 24–25
Disabling Injury Frequency Rate
    (DIFR), 17–18, 23–24
Disabling Injury Severity Rate
    (DISR), 17, 18–19, 23–24

Distractions, 112
Distributions, 34
  of accidents, 36
  binomial, 63–64
  normal, 47–48
  probability, 43–50
  probability functions of, 37–41
  of sample means, 46, 48
  t, 48–50
Dockboards, 127
Drainage systems, 128
Drill presses, 128
Dusts, toxic, 159

E

Electrical devices, stationary, 157
Electrical faults, 101
Electrical installations, 128
Emergency flushing, 129
Emissions, mean daily, 49
Emotional stability, 82
Environmental Protection Agency
    (EPA), 2
Equality, 13
Equipment
  automated, 101–102
  moving, 112
  personal protective, 145–146
  unsafe, 110–111
Error-commiting characteristics, 61
Error-provocation situations, 61
Errors
  in control systems, 99
  factors influencing, 61–62
  human, 100–101
  measurement, 15
  types I and II, 50–51
Event, 35–36
Exits, 129–130
Expected value, 41–42
Explosives, 130
Eye
  emergency flushing of, 129
  protection during welding, 161
Eye-hand coordination, 81

F

FAA, see Federal Aviation Agency

Factory inspection law, 2
Fair Labor Standards Act, 2
Fan blades, 131
FDA, see Food and Drug
    Administration
Federal Aviation Agency, 2
Federal Boat Safety Act, 2
Federal Hazardous Substances Act, 2
Federal safety agencies, 2
Fire department, first, 1
Fire extinguishers, portable, 131
Fire insurance company, first, 1
Fire protection, 131–132
Fire Research and Safety Act, 2
First aid, 144
Flammable liquids
    dip tanks for, 126–127
    drains for, 128
    incidental to principal business,
        132–134
Floors
    loading limit of, 134
    openings, hatchways, open sides of,
        135
    safety of, 134
Fluorine compound welding, 161
Fluxes, 161
Food and Drug Administration, 2
Foot protection, 136
Fork-lift trucks, 136
Franklin, Benjamin, 1
Frequency, 36

G

Gantry cranes, 121–122
Gases, toxic, 159–160
Girmaldi and Simonds method, 26–29
Guards
    point-of-operation, 148
    safety of, 137

H

Hand tools, 137–138
Hazardous conditions
    classification of, 178–180
    general, 136–137

Hazards
    automation-related, 99–101
    definition of, 8
    elimination of, 11
Head protection, 137
Heinrich, H. W., 2, 61
Heinrich method, 26
Highway Safety Act, 2
Hoists, 124–125
Housekeeping safety, 138
Human error, 8, 100–101
Hydraulic faults, 101
Hydrocarbons, halogenated, 122–123
Hypothesis, 50–55, 93

I

Illness rates, 21
Impact noise exposure, 145
Impulsive noise exposure, 145
Incidence, accident, 92–93
Indirect manual work, 98
Industrial Safety Activities
    measurement, 82–83
Industry safety and health checklist,
    115–163
Information system, 87–94
Injury
    body part affected by, 181–182
    classification by source, 175–176
    cost measurements for, 25–30
    nature of, 180–181
    rates of, 21–23
Interval scale, 13
Ionizing radiation exposure, 148–149

J

Job hazard analysis, 80
Job placement, 81
Jointers, 138
Journals, safety-related, 3–4

K

Knowledge and skill requirements
    (KSR), 98
Kolmogrov-Smirnov test, 44
Kruskal-Wallis Test, 78

L

Ladders, fixed, portable, 139–142
Lead welding, 161
Legislation, safety-related, 2
Lifting, unsafe, 111
Lighting, 142
Loading docks, 158
Lower control limit (LCL) values, 56, 71–72
Lunchroom safety, 142–143

M

Machines
   fixed, 143
   guarding for, 143
   working envelop of, 99–100
Masks, protective, 1
Mats, insulated, 143
Measurement
   definition of, 13
   indices of, 15–24
   in safety performance, 14–15
   scales of, 13–14
   techniques for, 16
Mechanical hazards, 101
Mechanical power transmission, 146–147
Mechanical power-transmission equipment, 123
Medical services, 145
Mental condition, 163
Mercury welding, 161
Metal coatings, 161
Mills, 121
Mine ventilation, 1
Mists, toxic, 159–160

N

National Electrical Code (NFPA 70–1971; ANSI CI-1971), 128
National Highway Transportation Safety and Administration (NHTSA), 2
National Traffic and Motor Vehicle Safety Act, 2

National Transportation Safety Board (NTSB), 2
Noise exposure, 144–145
Nominal scale, 13
Normal distribution, 44–46
Normal distribution function, 106–107
Nuclear Regulatory Commission (NRC), 2
Null hypothesis, 50, 53–54

O

Occupational health/safety, father of, 1
Occupational injury/illness, 21–23
Occupational Safety and Health Act, 2
   Form No. 200 of, 87
Occupational Safety and Health Administration (OSHA), 2, 20–21
Optimal cost-benefit model, 29–30
Ordinal scale, 13
Organizations, safety, 6
Oth machine's characteristics, 15
Overloading, 111
Oxygen cylinders, 126

P

Pareto's principle, 72
Passageway safety, 119–120
Personal protective equipment, 145–146
Physical characteristics, 14–15
Physical condition, 163
Physical effort, 98
Physical environment effect, 15
Placing, unsafe, 111
Pliny, the Elder, 1
Pneumatic faults, 101
Poisson distribution, 43–44, 55
   accidents calculated using, 58–59
   graph of, 45
   tables for, 105–106
Population distributions, 47–48
Position, unsafe, 111
Postures, unsafe, 111
Power-control devices, 162

Power transmission, mechanical, 146–147
Pressure vessels, portable unfired, 147
Prevention of Accidents in Coal Mines, Society for, 1
Probability, 36–37
  distributions of, 43–50
Probability density function, 37
  continuous, 39–41
  discrete, 37–39
  values of, 41–42
Programmer errors, 100
Protective clothing, 112–113
Protective devices, 112–113
Punch presses, 148
Pure Food and Drug Act, 2

R

Radial saws, 152–153
Radiation safety, 148–149
Railings, 149–150
  on fixed industrial stairs, 157
  on scaffolds, 154
Ramazzini, Bernadino, 1
Random variables, 34
  characteristics of, 41–43
  continuous, 34, 39–41
  discrete, 34
  probability density function of, 37–41
  variance of, 42–43
Ratio scale, 13–14
Refrigerator Safety Act, 2
Reliability, 13
Revolving drums, 150
Rolls, 121
Ropes, 121–122
Rubbish, 160

S

Safeguard, 8
Safety
  adequate knowledge of, 81
  assessment of, 8
  in automated plants, 98–102
  automation and, 95–102

awareness of, 81
  books on, 4–6
  data sources on, 7
  definition of, 1, 8
  enforcement of rules for, 80
  history of, 1–3
  information sources on, 88
  journals on, 3–4
  measurement indices for, 15–24
  organizations for, 6
  terms related to, 7–9
  workers' participation in, 81
Safety and Health Standard 70.a, 159–160
Safety appraisal scores, 78
Safety behavior
  control chart for, 71–72
  evaluation of improvement in, 80
  improvement of, 72, 75
Safety behavior sampling, 61
  correlated work in, 69–71
  fundamentals of, 62–64
  number of observations in, 67–68
  observer training for, 67
  pilot study in, 65–68
  procedure for, 65–72
Safety devices, inoperative, 110
Safety function, 8
Safety Information System (SIS), 87–89
  accident statistics in, 90–94
  computer package of, 165–171
  data base for, 89
  use manual of, 171–174
Safety inspections, 76–77
  checklist for, 115–163
  statistical inference of, 78–80
Safety lamp, 1
Safety measures, limitations of, 23–24
Safety performance
  composite score for, 82–84
  criterion for, 11–13
    measurement of, 13–16
    measurement techniques for, 16
    measures used for, 16–30
    need for, 12
  supervisor's, 80–81, 163
Safety training programs, 75

Sample, 8, 46, 48
Sample space, 35–36
Sampling, 35, 61–72
Sanitary conditions, 160
Saws
  band, 150
  portable circular, 151
  radial, 151–152
  swing or sliding cutoff, 152–153
  table, 153
Scaffolds, 154–155
SIS, see Safety Information System
Skills, 15
Sliding cutoff saws, 152–153
Social environment effect, 15
Social interaction, 98
Spray booths, 155–157
Spray-finishing operation, 155–157
Spraying areas, 155–157
Sprinkler systems, 132
Stairs, fixed industrial, 157
Standard, 8
Stationary electrical devices, 157
Statistical analysis, 33–58, 90–94
Statistics
  elements of, 33
  trends in safety inspections, 78–80
Stepladders, 140–141
Storage safety, 158
Supervisor, safety performance of,
    163
Swing saws, 152–153
System safety, 8
"System Safety Engineering for the
    Development of United States
    Air Force Ballistic Missiles,"
    2

T

t distribution, 48–50, 107–108
Table saws, 153
Tanks, open-surface, 158
Test statistics, 51–54
Toeboards, 158–159
  on open-sided platforms, 149
  on scaffolds, 155
Toxic Substances Control Act, 2

Toxic vapors, 159–160
Trash, 160
t-statistics, 52–54

U

U-bolt wire rope clips, 122
Uninsured costs measurement, 26–28
Union Fire Company of Philadelphia,
    1
Univariate distribution, 91–92
Unsafe acts
  automation and, 99
  classification of, 183–184
  definition of, 8
  list of, 65, 109–113
  measuring technique for, 61–72
Unsafe conditions, 61, 75
  automation and, 99
  definition of, 8
  detection of, 76–80
  features causing, 77
Unsafe speed, 110
Upper control limit (UCL) values,
    56–58, 71–72

V

V-belt drive, 128
Validity, 13
Variance, 42–43
Ventilation, 161

W

Walsh-Healy Public Contract Act, 2
Washing facilities, 160
Welding
  equipment safety in, 125–126
  safety in, 160–161
Woodworking
  belt sanders for, 120
  jointers for, 139
  machinery for, 161–162
  saws for, 150–153
Work pace, 98
Work requirements, 98
Work sampling, 62–72
Work stations, 65

Workers
  automation and performance of,
    95–98
  mental condition of, 81–82, 163
  physical condition of, 82, 163
  safety performance of, 81
Working conditions

  automation level and, 98
  safe, 81

Z

Z test-statistics, 51–52
Zinc welding, 161

DATE DUE

	261-2500		Printed in USA

366057